5725

**STRAND PRICE
$5.00**

D1713353

Lessons from Systems Thinkers

Problem-Solving and Analytical Thinking Methods from the Greatest Innovative Minds

By Albert Rutherford

Copyright © 2022 Albert Rutherford. All rights reserved.

All rights reserved. No part of this publication may be reproduced, distributed, or transmitted in any form or by any means, including photocopying, recording, or other electronic or mechanical methods, without the prior written permission of the publisher, except in the case of brief quotations embodied in critical reviews and certain other noncommercial uses permitted by copyright law. For permission requests, contact the author.

Limit of Liability/ Disclaimer of Warranty: The author makes no representations or warranties with respect to the accuracy or completeness of the contents of this work and specifically disclaims all warranties, including without limitation warranties of fitness for a particular purpose. No warranty may be created or extended by sales or promotional materials. The advice contained herein may not be suitable for everyone. This work is sold with the understanding that the author is not engaged in rendering medical, legal or other professional advice or services. If professional assistance is required, the services of a competent professional person should be sought. The author shall not be liable for damages arising therefrom.

The fact that an individual, organization of website is referred to in this work as a citation and/or potential source of further information does not mean that the author endorses the information the individual, organization to website may provide or recommendations they/it may make. Further, readers should be aware that Internet websites listed in this work might have changed or disappeared between when this work was written and when it is read.

First Printing, 2022.

Printed in the United States of America

Published by Kindle Direct Publishing

Email: albertrutherfordbooks@gmail.com

Website: www.albertrutherford.com

I have a gift for you…

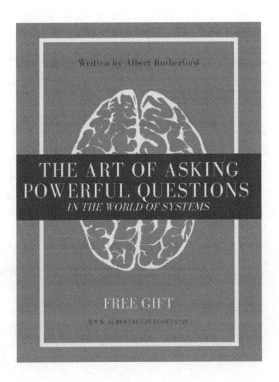

If you wish to receive notifications from me about:

- new book releases,

- new cognitive discoveries I made,

- book recommendations on how to develop your thinking toolkit further,

- ideas I'm pondering on,

Visit www.albertrutherford.com and fill in the subscribe box.

Thank you! Talk to You soon.

Table of Contents

I Have A Gift For You… 7
Table Of Contents 9
Introduction 10
Norbert Wiener 13
Warren Mcculloch 21
Gregory Bateson 27
W. Edwards Deming 35
Margaret Mead 43
Karl Ludwig Von Bertalanffy 51
Jay W. Forrester 61
Peter M. Senge 75
Professor Russell L. Ackoff 87
Peter B. Checkland 97
Ervin László 105
References 111
Endnotes 123

Introduction

Albert Einstein once asked this question, "have you ever thought about how you think?" A lot of people would say "no." It's not common for one to pause and think about their thoughts. Herein lies the problem. What Einstein pointed out with his question is that certain issues cannot be solved with the same kind of thinking that created those issues. Second - and third-guessing every thought crossing our mind is not a reasonable expectation. It is time-consuming and, frankly, tiring. What we can do, however, is learn new thought patterns and use them to our benefit. The rewired thinking path I'm talking about is called systems thinking. My mission in this book is to present you the evolution of this cognitive assessment style through the stories of the people who asked hard questions and developed better solutions to the problems they were facing.

Today's scientific exploration has come a long way from where it used to be. In the past, science tended to look at events as individual occurrences that seemingly happened in isolation. Different science fields would concentrate on the event level without seeing how to fit it into the bigger picture. It is strange to carefully study, examine, and report on what you had observed about one puzzle piece without ever looking at the whole puzzle, isn't it? Well, pre-modern science functioned similarly.

Different fields of science have evolved throughout the years, and there has been a paradigm shift in data assessment. Scientists are still concerned with carefully examining individual events, but now looking at how it fits into the whole picture is also relevant. Things are more often interdependent than independent. Many scientific fields shifted from deconstructing elements to their particles to analyzing their dynamics and working-togetherness.

General System Theory -and later systems thinking - didn't want to abandon the foundation of knowledge of any field of science. It aimed to be transdisciplinary. Each field would bring its expertise to the table. They would actively engage in working together to create a common base of methods and for all scientific disciplines to share.

General Systems Theory wanted to "expand the tent;" to be more inclusive of knowledge from all areas. To stop being so focused on the little pieces that seeing the bigger picture becomes impossible.

Today systems thinking – while its methodology has changed and refined throughout the years - is actively used in politics, economics, sociology, demographic analysis, and environmental studies. This book will tell the story of how systems thinking as a field of study came into being and evolve throughout the decades. It will present the thoughts and contributions of eleven great minds whose ideas were vital to develop this critical thinking method.

Norbert Wiener

Norbert Wiener was an American mathematician who attained renown due to formulating some of the most significant contributions to mathematics in the 20th century. His expansion of the field led to the development of cybernetics, which examines the interaction of feedback loops and behavior.

Born on November 26, 1894, in Columbia, Missouri, he showed a gift for mathematics early and graduated from high school at the age of 11.[i] He then attended Tufts College and graduated with a BA in Mathematics in 1909 at the age of 14. After spending a year at Harvard as a graduate student in zoology, he realized laboratory work was not his forte and transferred to a graduate degree in Philosophy. At 17, he graduated from Cornell with his graduate degree and returned to Harvard to pursue a Ph.D. He earned his degree at the age of 19 after successfully defending his dissertation on mathematical logic. He traveled first to England to study mathematical logic at the University of Cambridge and then to the University of Göttingen in Germany to study differential equations, both on a grant from Harvard. He published his first paper in the mathematical journal *Messenger of Mathematics* at Cambridge in 1913.[ii]

During World War I, he could not enlist due to his poor eyesight. He tried a variety of jobs over the following five years, including being a teacher at the University of Maine, a writer for an encyclopedia, an apprentice engineer for General Electric, a journalist for the Boston Herald, and a mathematician in the Aberdeen Proving Grounds in Maryland. In 1919, he was hired as an instructor for the mathematics department at the Massachusetts Institute of Technology (MIT) around the time it started building towards being a center for enhanced learning in science and technology. He remained at MIT on the faculty until his retirement. During the 1920s, he was involved with work on what is now referred to as stochastic processes and the theory of Brownian motion and generalized harmonic analysis. [iii]

During World War II, Wiener worked on issues surrounding aiming guns at a moving target. This experience led to the creation of "Extrapolation, Interpolation, and Smoothing of Stationary Time Series," first appearing as a classified report. Wiener became a co-discoverer of the theory of the prediction of stationary time series. This work also allowed him to develop the concept of cybernetics and the Wiener Filter.[iv]

Weiner published his book *Cybernetics; or, Control and Communication in the Animal and the Machine* in 1948. His popular scientific book helped him become renowned in the scientific community. Wiener continued working at cybernetics and philosophized about it all his life while maintaining research projects in other areas of mathematics.[v]

After the war, he continued to contribute new ideas to various subjects, including mathematical prediction theory and quantum theory, showing how the latter is consistent with other branches of science by applying his theoretical description of Brownian motion to quantum phenomena.

Wiener wrote many works throughout his life. These include *The Human Use of Human Beings*, published in 1954, and *God and Golem, Inc.: A Comment on Certain Points Where Cybernetics Impinges on Religion*, published in 1964, and two volumes of autobiography, *Ex-Prodigy*, published in 1953, and *I Am a Mathematician*, published in 1965. In 1933, he won the Bocher Memorial Prize for his work on the foundations of calculus theory. Then, in

1963, he was awarded the National Medal of Science, receiving the medal the following year, only a few weeks before his death.[vi]

Why is he essential to systems thinking?

Wiener played an important role in developing systems theory, which took root during World War II when he worked on anti-aircraft fire issues. He was inspired by these studies surrounding communication and control of specific technical systems, which led him to consider more general concepts of what he came to call the science of cybernetics. His primary argument concerning systems was that any system could be understood when general principles are applied. When examining Wiener's cybernetics, it involves two paired ideas. The first includes:

1) Feedback, which has regulating properties and stabilizes the various interactive parts.

2) Transmission of information, which assists in transforming the various, unrelated parts of a complex system and brings them together as a whole.[vii]

The term "feedback" can be used to express the relationship between control and communication. This relationship is centered on the fundamental notion of the "message" that describes a continuous sequence of measurable events throughout time. An example is an automatic controller that maintains a constant temperature in a room by continuously monitoring any outside influences on the temperature.[viii]

The concept of the message was vital to developing the communications theory and Wiener's definition of "information" as a measurement of the degree of organization within a system.

Negative feedback is viewed as more useful than positive feedback as it gives the appearance of stability and control, whereas positive feedback can be viewed as dangerous and unstable. An example of positive and negative feedback can be seen in a procrastinating medical student. If the student has an upcoming exam and does not adequately prepare due to procrastination, he could fail the exam. This would result in negative feedback and cause a change in his future behavior, i.e., he will be more likely to prepare for upcoming exams properly and therefore learn the material needed to be an effective doctor. If, however, he procrastinates on studying and still passes the exam, he will likely continue to procrastinate when studying for future exams. This will result in positive feedback, where it is reinforced that he will pass exams even if he is not adequately prepared for them, which could cause harm to his future patients.

On a broader scale, control and communication create an organization that nature will inherently try to destroy due to entropy. If information is deprived of organization and meaning, as often happens due to nature's tendency to not care about human rationality, this leads to limitations of communication within and among individuals. Humans perceive information from the world through their sense organs that then travels to and is coordinated in the brain before emerging through effector organs, such as the muscles, if an action is warranted.[ix] Perceiving and storing information will influence future actions. This exchange of information from the external world allows humans to adjust and thrive in their environment, which continues to grow more complex as time passes.

The second pair of concepts involved with the definition of cybernetics are:

1) Humans and animals and their physiology.

2) Machines and their mechanics.

Wiener conducted two major research projects in the 1930s to 1940s, one examining feedback within human and animal physiology, and the other was the previously mentioned study on building control systems for anti-aircraft weaponry, which was also based on feedback principles. The results of these two studies can be combined, explaining how all purposeful behavior may require negative feedback.[x] This parallel between human and machine activities shows the importance of, for example, the early development of digital computers. As described in later studies, humans were viewed as information-processing entities similar to intelligent machines. Equating activities between humans and machines proved to be an important concept in multiple disciplines, such as computing and psychology. The field of cybernetics helped make this parallel widely known.

Warren McCulloch

"Don't bite my finger, look where I'm pointing!"

– Warren McCulloch (quoted by Seymour Papert).

Warren McCulloch had many identities in his scientific life, including "philosopher, poet, neurologist, neurophysiologist, neuropsychiatrist, collaborator, theorist, cybernetician, mentor, and engineer."[xi] While Weiner coined the term "cybernetics," Warren McCulloch organized and chaired the Macy conferences, published the first logical model of the mind, and supported and mentored the upcoming key figures in cybernetics. Yet, many people who followed in his legacy do not even know his name, as if he has faded from view.[xii]

McCulloch was born on November 16, 1898 in Orange, N.J. McCulloch studied at Yale, where he received a B.A. in philosophy and psychology in 1921, a Master of Arts in psychology in 1927 from

Columbia, and an M.D. from New York's College of Physicians and Surgeons in 1927.[xiii] After graduating, McCulloch began research in experimental neurology and examined areas of the brain involved with epileptic seizures. He then spent several years working at Bellevue Hospital in Manhattan and at the Rockland State Hospital for the Insane before returning to Yale in 1934. There, he worked in a neurophysiology laboratory evaluating the brain's functional organization.[xiv]

From 1935 until 1941, McCulloch worked stints at Yale, first as a Sterling Fellow, then an instructor, and finally as an assistant professor. In 1941, he relocated to the University of Illinois College of Medicine and worked there for the next seven years. During this time, he formed collaborations to examine the relations between epilepsy and schizophrenia, researching the mechanisms behind insulin shock therapy as a treatment of schizophrenia, based on the incorrect belief people with the diagnosis never developed epilepsy. In 1943, he coauthored his most famous paper, "A Logical Calculus Immanent in Nervous Activity." He also collaborated on the logic of the central nervous system, presenting a hypothetical model in 1947 of the functional relations between neurons that achieved recognition of universal forms. In 1952, he said goodbye to Illinois and went to work at MIT, spending the rest of his life examining brain function within the framework of information processing. In 1964, he was declared an honorary founder and first elected president of the American Society for Cybernetics in recognition of his foundational work in the field and chairing the Macy conferences. He died in Cambridge, Massachusetts in 1969.[xv]

Why is McCulloch important in systems thinking?

During McCulloch's time as a research psychiatrist in the 1940s, he became an important figure of the American cybernetics movement. His form of cybernetics used logic and mathematics to develop models of neural networks, describing the function of the mind as part of the working brain. As mentioned, he collaborated in 1947 with Walter Pitts on the hypothetical model of the functional relations between neurons, such as the perception of the form "triangle." This led to the creation of a model of the visual cortex that involved a complex array of neuron-carrying impulses, ultimately leading to a projected image in our brains we can interpret and identify. Their model proved too mathematical without showing reality accurately.[xvi]

An earlier collaboration between the two led to a paper published in 1943 called "A Logical Calculus of the Ideas Immanent in Nervous Activity." In this paper, they created a computation neural model called McCulloch and Pitt's model of a neuron (MCP), based on mathematics and algorithms called threshold logic. They were attempting to understand how the brain could produce highly complex patterns by using connected neurons.[xvii] The MCP was fairly simple, and this posed challenges. It could only receive binary inputs and outputs and lacked an ability for learning. Nevertheless, the MCT led to two major neural network research approaches; one focused on biological processes in the brain while the other examined the application of neural networks to artificial intelligence.

McCulloch and Norbert Wiener never worked directly together. However, Wiener influenced McCulloch's thinking about cybernetics. That being said, McCulloch had his unique and independent contribution to the field. Their main difference regarding feedback, hearing devices, and perception was that Wiener concerned himself with truth and normality while McCulloch explored the possibilities of knowledge generation.

McCulloch was a paradox with a fascinating mix of ideas. He self-described as an aspiring theologian before becoming interested in mathematics and its problems surrounding science. His pursuit of understanding the workings of the brain led him through his various degrees. He describes having bright colleagues and pupils, who frequently came up with new, inspiring ideas, stimulating him all the time.

McCulloch had a firm belief that, to promote a foundation framework for psychiatry, it had to be rooted in the science of the brain. However, his developed models were often criticized as being too speculative and unable to incorporate sufficient empiric data. Still, he was consistently judged by his peers as brilliant, inspiring new theoretically-minded brain researchers and computationalists to continue expanding the field.[xviii]

Gregory Bateson

Gregory Bateson was an anthropologist, linguist, philosopher, and social scientist.[xix] He made significant contributions to different sciences, including anthropology, psychiatry, and the new interdisciplinary field of cognitive science, which he helped pioneer.[xx] Furthermore, he utilized systems theory in social sciences. This mix led to the identification of the paradox of the double bind. I will describe this shortly. Overall, Bateson developed a new way of thinking in terms of relationships, connections, patterns, and context.[xxi]

He was born on May 9, 1904 in Grantchester, England. He began his studies at the Charterhouse School and then received his bachelor's in biology from St. John's College. He continued his studies at St. John's at Cambridge. In 1929, he decided to get as far away from Mother England as possible, accepting a linguistics teacher position in Australia at the University of Sydney.[xxii] Bateson was fascinated with the work of Charles Darwin. He pursued his passion in anthropology, repeating the fieldwork Darwin completed 100 years prior. Despite his adventures, he didn't stay with the Aussies for long. In 1931, he returned to Cambridge University as a Fellow.

A curiosity in Bateson's life is his marriage to a renowned anthropologist and systems thinker, Margaret Mead, in 1936. (She has her own chapter in this book.) The couple stayed together for 14 years, contributing and influencing each other's work.

The Double Bind

The concept for which Bateson is most well-known is the paradox of the double bind. This concept presented a dilemma while he researched schizophrenia with his colleagues. What is a double bind? It happens when an individual experiences conflicting emotional, verbal, or physical messages.[xxiii] People with emotional impairment had trouble processing the conflicting signals they were receiving from the external world and from within. Later, Bateson and his team defined that the double bind can be a tool for thought control. People can communicate assumptions not only verbally but also using intonation, gestures, or eye contact. Those who experience intimidation due to double bind can feel trapped in uncomfortable situations, where they are expected to do things that may result in a positive or negative outcome. An example of this can be seen in an abused individual who is told they are loved while being informed they will no longer be loved if they tell anyone about the abuse. The double bind can be utilized as a coercion and control technique in relationships, with one party gaining power over the other. This can elicit feelings of fear and powerlessness and the denial of a part of reality in the controlled party. Initially, Bateson utilized the framework for the double bind to understand people suffering from schizophrenia. As his theory progressed, he broadened this concept, creating a systems approach, applying the double bind in family therapy.[xxiv] The major criticism of Bateson's double bind approach to schizophrenia was that it put an unfair blame on the parents for their child's mental condition. Schizophrenia is a physical disorder of the brain that is hard to be prevented by parenting.[xxv]

Why is he important in systems thinking?

Bateson was an original thinker who crossed multiple disciplines. His intellectual journey takes him from biology to anthropology, to pathologies of systems of ideas, to systems of ideas involving how we all live together, including plants and animals. His various works laid the foundation for the expansions in cybernetics that followed and the ever-evolving problems in today's world. He attended all ten Macy conferences in the series, powerfully influencing the meetings as the members delved into the concepts surrounding cybernetics. He later talked about the importance of the Macy conferences, with references to cybernetics appearing in his later works.

His contributions to our knowledge today are difficult to quantify, especially as they cross multiple disciplines. Here are only a handful of examples of Bateson's ideas:

1) Schismogenesis, an anthropological concept. This term describes a positive feedback loop that allows for the increasing destruction of relationships.[xxvi] This term can further be divided into two types:

> a) "Complementary schismogenesis: An example of this can be characterized by a class struggle between two groups of people, where one side elicits behavior X while the other side elicits behavior Y. These two behaviors complement one another, such as in the dominant-submissive behaviors of a class struggle. These behaviors may also

exaggerate each other, causing a rift and possible conflict.

b) Symmetrical schismogenesis: An example of this can be characterized by an arms race, where the behaviors of the parties influence similar behaviors from the other parties. When examining the United States and the Soviet Union, each party would continually try to amass more nuclear weapons than the other party."[xxvii]

2) Double bind, a psychotherapeutic concept. This term describes the patterns of interaction between people who must interact incompatibly with each other simultaneously.[xxviii] An example was described earlier.

3) Levels of learning. This term describes how some forms of learning are at a higher logical level than others, which creates many ways to learn how to learn.[xxix] These levels are divided into Learning 0 to Learning 4, although he didn't discuss Learning 4. Learning 0-3 are described below:

a) Learning 0: the most basic level of learning, which only entails responding to stimuli, with no experience or information-backed changes.[xxx]

b) Learning 1: it is the ability to optimize alternative choices within a pre-determined set of alternatives. It is based on past experiences, influenced by one's mental models and specific goals, ending in outcomes by taking specific alternatives. The result is knowledge about the optimal choice of alternatives within the list of the given alternatives.[xxxi]

c)   Learning 2: it is a revision of Learning 1, where the set of alternatives becomes dynamic. It is the combination of past knowledge coupled with knowledge about the new dynamic alternatives. [xxxii]

d)   Learning 3: it is a revision of Learning 2, where there is a corrective change in the system of sets of alternatives from which the choice was made. The focus is on the underlying mental model linked to one's needs and values. The outcomes involve creating knowledge of the current mental model when selecting alternatives and becoming aware of the unconscious elements of that mental model.[xxxiii]

Additional ideas can be added to these concepts, including Bateson's definition of "the difference that makes a difference" and the phrase he borrowed from Alfred Koryzbski, "the map is not the territory."[xxxiv] Bateson's ideas had a significant impact on society, for instance, how the concept of double bind played a foundational role in developing family therapy, and the levels of learning have contributed to organizational learning.[xxxv] It was towards the end of his life that Bateson realized how closely linked all his ideas were and how they encompassed a way of thinking that could be used systematically across multiple subject matters.

W. Edwards Deming

Born on October 14, 1900, in Sioux City, Iowa, W. Edwards Deming came from a modest background. His family needed to be frugal and didn't have much to waste, thus Deming grew to be an adult with a keen sense of good wealth management. Despite humble beginnings, Deming was quite studious, earning an engineering degree at the University of Wyoming in 1921 and a Master's in mathematics and physics from the University of Colorado in 1924. After being awarded a doctorate in mathematical physics from Yale University in 1928, he spent the next ten years lecturing and writing in mathematics, physics, and statistics and working as a mathematical physicist at the United States Department of Agriculture from 1927 to 1939. Afterward, he worked as a statistical adviser for the U.S. Census Bureau from 1939 to 1945 then as a business consultant and a professor of statistics at New York University's graduate school of business administration from 1946 until his death in 1993. He was also appointed as a distinguished professor in management at Columbia University in 1986. [xxxvi]

Deming became interested in statistical analysis in the 1930s. He wanted to improve methodologies used to achieve better quality control in compiling and comprehending the life of product defects, their identification, and the analysis of root causes. After the defects were corrected, he monitored the results to determine the effects of those corrections on subsequent product quality. In 1950, he was invited to Japan to teach executives and engineers his techniques. As a result, many Japanese companies committed to Deming's quality control methods, allowing them to dominate areas of the world market throughout the world. Japanese quality was renowned before the training, but especially ever since. American corporations were relatively late adopters of Deming's methods, embracing them in the 1980s.[xxxvii]

Deming is the author of several books and roughly 170 papers. Some examples include *Quality, Productivity and Competitive Position*, published in 1982, "Out of the Crisis," posted in 1986, and *The New Economics*, published in 1993. He also offered four-day seminars that had an attendance of around 10,000 people per year for over ten years.[xxxviii] He received many awards and recognitions throughout his life due his many contributions worldwide. These include the annual Deming Prize, which Japanese manufacturers created in his honor. He was also decorated with the Second Order Medal of the Sacred Treasure in May 1960 by the Emperor of Japan. In 1986, he was elected to the National Academy of Engineering and the Science Technology Hall of Fame in Dayton. He was also awarded the Distinguished Career in Science from the National Academy of Sciences in 1988.[xxxix]

Why is he important in systems thinking?

Deming's ideas on quality stem from the recognition of the importance of variation. In a quote from "Out of the crisis," he indicates:

"The central problem in management and in leadership… is failure to understand the information in variation." – Deming

Systems throughout the world have variation – at least in their elements. However, Deming believed it was important for managers to see the difference between common and special variation causes. Deming's theory of variation suggests special cases of variation can be attributed as side effects from easily identifiable factors, including changes in procedure or change of shift or operator. However, common causes will continue to exist even with the removal of special causes. The workers usually recognize common causes; however, it requires authority from the managers to implement the changes needed to prevent the same problem from resurfacing. In Deming's estimate, about 85% of the causes of variation were due to mismanagement.[xl]

Deming's expertise in management and quality control led him to develop the famous "14 Points for Management" that can also be found in "Out of the Crisis." These points include:

1) "Create constancy of purpose toward improvement of product and service, aiming to become competitive, stay in business, and provide jobs.

2) Adopt the new philosophy. We are in a new economic age. Western management must awaken to the challenge, learn their responsibilities, and take on leadership for change.

3) Cease dependence on inspection to achieve quality. Eliminate the need for inspection on a mass basis by building quality into the product in the first place.

4) End the practice of awarding business based on the price tag. Instead, minimize total cost. Move toward a single supplier for any one item, on a long-term relationship of loyalty and trust.

5)   Improve constantly and forever the system of production and service, to improve quality and productivity, and thus continuously decrease costs.

6)   Institute training on the job.

7)   Institute leadership. The aim of supervision should be to help people and machines and gadgets to do a better job. Supervision of management is in need of overhaul and supervision of production workers.

8)   Drive out fear, so that everyone may work effectively for the company.

9)   Break down barriers between departments. People in research, design, sales, and production must work as a team, to foresee problems of production and in use that may be encountered with the product or service.

10)   Eliminate slogans, exhortations, and targets for the workforce asking for zero defects and new productivity levels. Such exhortations only create adversarial relationships, as the bulk of the causes of low quality and low productivity belong to the system and thus lie beyond the power of the workforce.

a)   Eliminate work standards (quotas) on the factory floor—substitute leadership.

b) Eliminate management by objective. Eliminate management by numbers and numerical goals. Substitute leadership.

11) Remove barriers that rob the hourly worker of his right to pride in workmanship. The responsibility of supervisors must be changed from sheer numbers to quality.

12) Remove barriers that rob people in management and the engineering of their right to pride in workmanship. This means, inter alia, abolishing the annual or merit rating and management by objective.

13) Institute a vigorous program of education and self-improvement.

14) Put everybody in the company to work to accomplish the transformation. The transformation is everybody's job."[xli]

Deming went on to describe the main barriers he perceived management would face when attempting to improve effectiveness. They are referred to as the "Seven Deadly Diseases of Management":

1) "Lack of constancy of purpose to plan products and services that will have a market and keep the company afloat.

2) An emphasis on short-term profits and short-term thinking (just the opposite from constancy of purpose to stay in business), fed by fear of unfriendly takeover, and by demand from bankers and owners for dividends.

3) Evaluation of performance and annual reviews.

4) Mobility of managers.

5) Management by use only of available data.

6) High medical costs.

7) High costs of liability."[xlii]

On top of a commitment to quality to combat these deadly diseases, Deming emphasized the importance of communicating quality messages to all staff members and promoting faith in total quality management. These principles can be applied to wide general management, which led to Deming being credited as the founder of the Quality Management movement.[xliii]

The PDCA (Plan-Do-Check-Act) Cycle was originally a concept developed by Walter Shewhart, who introduced it to Deming. It later became known as the Deming Wheel or Cycle in the 1950s due to Deming's promotion of the idea. The cycle consists of four steps to assist in facing a problem and successfully resolving it. These steps include:

1) "Plan for changes to bring about improvement.

2) Do changes on a small scale first to trail them.

3) Check to see if changes are working and investigate selected processes.

4) Act to get the greatest benefit from the change."[xliv]

The originality of Deming's philosophy came not from the world of management, but from mathematics. He also mixed in an observation-based human relations approach instead of a specific management theory. He keenly analyzed people's needs in their working environment to allow them to live up to their work performance potential.

Margaret Mead

"A small group of thoughtful people could change the world. Indeed, it's the only thing that ever has." – Margaret Mead.

Margaret Mead was an American cultural anthropologist and writer, born on December 16, 1901 in Philadelphia, Pennsylvania. Her parents were social scientists, who valued education and social issues. Their principles influenced young Mead and offered direction in later life and career. She grew up in the Progressive Era, when reformers trusted that social problems could be addressed with the application of social sciences. This resulted in criticism in her later years due to her belief that traditional cultures should adopt Western ways to allow their society to progress.[xlv]

In her earlier years, she and her siblings were encouraged to spend more time out of the house, learning outdoors and experiencing practical lessons in natural history and botany. Her paternal grandmother, a widowed schoolteacher who lived with Mead's family, played an active role in the children's education and believed it was harmful to children to be indoors for long periods of time. When Mead's two youngest sisters were born, her grandmother also encouraged her to take notes on their behaviors while they were babies. This allowed her to witness the differences in temperament that developed between the two girls. Mead's upbringing promoted her to develop as a keen observer of the world around her, starting in her early years.[xlvi]

Mead began her educational career at DePauw University in 1919, but after a disappointing experience, she transferred to Barnard College the following year. There, she began a degree in English but later switched to psychology. She started taking classes in anthropology. This is where she met Franz Boas and his teaching assistant, Ruth Benedict, the former being often referred to as the "father of modern American anthropology." As Mead explored primitive cultures in her classes, she observed their potential avenue for exploring an important question in American life: "How much of human behavior is universal, therefore natural and unaltered, and how much is it socially induced?" Answers to these questions could have important social consequences, especially in terms of gender roles and among those who deem women to be inferior to men.

Mead's work with Boas and Benedict influenced her to become an anthropologist. Boas' was tasked with documenting cultures before they disappeared following contact with the modern world (called "salvage anthropology"), which fueled Mead's interest in the field. Benedict also became Mead's mentor and long-term colleague, encouraging her to focus on anthropology and all she could offer in the field. She later received her M.A. in 1924 and her Ph.D. in 1929 at Columbia University.[xlvii]

Her doctoral dissertation was a study of the relative stability in certain elements of culture, focusing on canoe building, house building, and tattooing in five Polynesia cultures, including Samoa.[xlviii] Her basic study was completed in 1925 before traveling to Samoa. She had many expeditions to Samoa and New Guinea, exploring the adaptability of human nature and the variability of social customs. Her first book was published in 1928, titled *Coming of Age in Samoa*, where she described the easy transition of Samoan children into the sexuality and work of adulthood and how this contrasted with the restrictions on sexual behavior and separation from the productive world experienced by children in the United States.[xlix] It became an instant bestseller and remains in print today. Her second book, *Growing Up in New Guinea*, was published in 1930. She made 24 field trips altogether among six groups of South Pacific populations.

Mead continued exploring the Westerners' troubles surrounding sexuality in her later book *Sex and Temperament in Three Primitive Societies*, published in 1935. In this book, she described the variety of temperaments among men and women in different cultures. In the Arapesh culture, men were nurturing, unlike the violent women of the Mundugumor tribe.[1] Mead stated social cultures determine human behavior, not biology, thus occupying a strong stance in the nature vs. nature debate. She stood for nurture. Mead observed that children learn by watching adult behavior – she called this phenomenon imprinting. She continued exploring her stance in her later works, *Male and Female: A Study of the Sexes in a Changing World*, published in 1949, and *Growth and Culture*, published in 1951. In these books, she described her belief that differing personality characteristics among men and women were shaped by cultural customs instead of heredity. She also analyzed the ways in which motherhood perpetuated the traditional roles of men and women in all societies and emphasized the possibility for individuals to resist traditional gender stereotypes.[li]

While she had been married three times, her third marriage was to Gregory Bateson, another anthropologist who was discussed in an earlier chapter. They collaborated on many works, such as working together in Bali from 1936 to 1938 and among the Iatmul of New Guinea from 1938 to 1939. They obtained a significant number of still photographs and film footage, allowing them to produce two photographic ethnographies and seven edited films, including *Balinese Character: A Photographic Analysis* in 1942. These greatly affected visual anthropology and the discipline overall a few years after publication.[lii]

Mead had only one child and was divorced three times. However, she was considered an expert on topics such as family life and child-rearing. She would often try to encourage Americans to understand the lives of other people as it could help them understand their own. Also, she emphasized that having a less rigid stance regarding sexuality, both heterosexuality and homosexuality, could enrich their lives, combining motherhood and careers is possible and should be actively sought, and creating support networks for the overwhelmed "nuclear family" would allow greater well-being for everyone involved.[liii]

Another notable fact about Mead is that she was appointed as an assistant curator of ethnology at the American Museum of Natural History in 1926. She successfully served until 1942 then became an associate curator until 1964, curator of ethnology until 1969, and curator emeritus until her death in

1978. She was elected at the age of 72 to the presidency of the American Association for the Advancement of Science due to her contributions to science. After her death, she was posthumously awarded the Presidential Medal of Freedom in 1979.[liv]

Why is she important in systems thinking?

"There has been an increased but still rather limited response to general systems theory, as variously reflected in the work of Bateson, Vayda, Rappaport, Adams, and an interest in the use of computers, programming, matrices, etc. But the interaction between general systems theory (as represented, for example, by the theoretical work of Von Bertalanffy) has been compromised, partly by the state of field data, extraordinarily incomparable as it inevitably is, as well as historical anthropological methods of dealing with wholes. General systems theory has taken its impetus from the excitement of discovering larger and larger contexts, on the one hand, and a kind of microprobing into fine detail within a system, on the other. Both of these activities are intrinsic to anthropology to the extent that field work in living societies has been the basic disciplinary method. It is no revelation to any field-experienced anthropologist that everything is related to everything else, or that whether the entire sociocultural setting can be studied in detail or not, it has to be known in general outline." — Margaret Mead, Changing Styles of Anthropological Work, 1973.

There was widespread awareness of Mead's work in her own time as well as today. However, she is generally identified with her discipline, cultural anthropology, not as a systems thinker. This is unusual, as she played a big role in the establishment of systems thinking and exhibited strong systemic awareness in her work. Elise Boulding, a Professor Emerita of Sociology, observed in her book published in 1995, "studying micro societies is an ideal way to get inside the dynamics of social process, and Mead made the most of her opportunities. She was always asking the questions: What makes things change? What makes things stay the same?"[lv]

Mead had significant involvement with the Macy conferences, being one of the core group of social scientists, along with Larry Frank and Bateson, who worked alongside the physical scientists in developing the mathematics of cybernetics. She was described as an influential spokesperson, acting as a conduit and champion of the role of social sciences in the conferences. She assisted with editing the proceedings of the sixth to the tenth conference, alongside Heinz von Foerster and Hans-Lukas

Teuber, and expressed a clear understanding that the world of social science was ready for the concepts of cybernetics. Her involvement in the organization of the systems movement continued after the conclusion of the Macy conferences. Mead was also an early leader of the Society for General Systems Research, founded by von Bertalanffy.[lvi]

Karl Ludwig von Bertalanffy

Karl Ludwig von Bertalanffy was an Austrian biologist, born on September 19, 1901. He is known as one of the founders of general systems theory (GST). Von Bertalanffy grew up in Atzgerdorf, a little village near Vienna, Austria. He was home schooled until he was 10 years old, educated by private tutors. Thus, he was versed in self-studying by the time he went to a regular school, being able to keep up with the lectures. The famous biologist Paul Kammerer lived in Von Bertalanffy's neighborhood and became his mentor at a young age.[lvii]

He began university-level studies in philosophy and art history in 1918, first at the University of Innsbruck and later at the University of Vienna. He got to a point where he had to decide whether to continue his studies in philosophy or move to biology. Young Von Bertalanffy chose biology, saying one can become a philosopher later but not a biologist.[lviii] In the 1920s, biology was still considered an experimental science. This didn't deter Von Bertalanffy. In fact, he developed much-needed quantitative models for theoretical biology. He was an early supporter of the "organismic" conception of biology and focused on the whole system of organisms. He was also involved with experimenting on the diagnosis of cancer by using cellular screening.[lix]

In 1934, von Bertalanffy was appointed Privatdozent of the University of Vienna, an academic title given to people with formal qualifications that prove the ability and grant the permission to teach a given subject at the highest level - yet gives little compensation.[lx]

Von Bertalanffy's wife, Maria, whom he met in April 1924, devoted her life to von Bertalanffy's career. She worked both for him and with him in his career after they moved to Canada and compiled two of his last works after his death. Their one son would also follow in his father's footsteps by entering the field of cancer research as his profession.[lxi]

Why is von Bertalanffy important in systems thinking?

"General system theory, therefore, is a general science of wholeness... The meaning of the somewhat mystical expression, "The whole is more that the sum of its parts" is simply that constitutive characteristics

are not explainable from the characteristics of the isolated parts. The characteristics of the complex, therefore, appear as new or emergent..." - Ludwig von Bertalanffy

General systems theory (GST) is one of von Bertalanffy's biggest contributions to the world. It is an interdisciplinary practice that describes systems with interacting elements. He attempted to give alternative options to conventional models of organization. In his GST, von Bertalanffy emphasizes holism and organisms over reductionism or "atomism" - the latter two theories characterized American Behaviorism and orthodox psychoanalysis.[lxii]

Reoccurring themes in von Bertalanffy's works highlight the importance of the system as a whole and the idea that what distinguishes a system from a collection of parts is that the system has a form of organization of its parts. Organismic biology looked at biological processes as explainable elements with the help of physics and chemistry. However, the theory of vitalism, which proposes a mysterious underlying force creating the complexity of life, was rejected by organismic biology. Von Bertalanffy deemed vitalism's explanations weak and closely connected to the reductionist worldview. Despite his disdain, he used some of Hans Driesch, a vitalist biologist's, concepts such as "equifinality," which states the same end point in the development of an organism may be reached in many ways from different starting points.[lxiii]

GST most likely arose from von Berfalanffy's ideas in biology. He looked at the organism as a whole and its openness to the surrounding environment. Thus, another well-known term he coined was the theory of "open systems." This theory can be applied to many fields, including biology, cybernetics, education, history, philosophy, psychiatry, psychology, and sociology. Von Bertalanffy used the general systems theory framework to reconcile the material scientific approach to human behaviors with the humanistic one.

In his theory of open systems, von Bertalanffy claimed the traditional closed systems models utilized in classical science were ill-founded for living organisms. He said the second law of thermodynamics[1] also fails to prove true with open systems. "The conventional formulation of physics are, in principle, inapplicable to the living organism being open system having steady state. We may well

---

[1] "The Second Law of Thermodynamics states that "in all energy exchanges, if no energy enters or leaves the system, the potential energy of the state will always be less than that of the initial state." This is also commonly referred to as entropy." From https://www2.estrellamountain.edu/faculty/farabee/biobk/biobookenerl.html

suspect that many characteristics of living systems which are paradoxical in view of the laws of physics are a consequence of this."[lxiv] The question whether open physical systems could justify a general systems theory remained to be seen.

Open systems are defined by negative entropy, a phenomenon that "explains organismic growth, differentiation, increasing complexity, goal directness, and purposefulness of human and animal behaviors."[lxv] An open system has a clear boundary that divides the inside of the system from the outside. However, matter and energy can still cross. This contrasts with a closed system, in which only energy can cross the border. Furthermore, an open system does not stay in a traditional state of equilibrium. Instead, it is constantly changing, even though it retains its basic form. Von Bertalanaffy referred to this situation as "dynamic equilibrium" or "steady state", a concept he developed in exact mathematical form and has been used by many other authors. The application of open systems to wider situations led von Bertalanffy to develop GST to determine principles that applied to systems in general, classify the different types of systems, and create mathematical models that could describe them, hoping to unify science.[lxvi]

GST and Cybernetics

Von Bertalanffy's was a humanist; he valued the role of ethics in science. He applied this principle to psychology to go against the dehumanizing "robot model" of behaviorist psychology, functionalistic sociology, and cybernetics.

Albeit GST and cybernetics share a few founders, such as the proponents of Gestalt theory and Jean Piaget, they are conceptually different. Von Bertalanaffy asserted that, while GST and cybernetics developed in parallel, they began at different starting points and basic models (basic science vs. technological applications, and dynamic interaction and open systems vs. feedback and hemostasis, respectively).[lxvii] Von Bertalanffy gave credit to cybernetics for its insights into regulatory, goal-seeking, and teleological behavior, which are often issues with GST. However, he also claimed cybernetics "falls short of being a general theory of systems or providing a new 'natural philosophy'."[lxviii]

Von Bertalanffy applied general systems concepts in social sciences, such as the concept of general system, of feedback mechanisms, interactions, and communication. The complexities of the interactions between natural sciences and human social systems, however, posed difficulties when trying to cover social sciences with a new general theory. Despite these hardships, von Bertalanffy believed the GST stimulated new developments in various scientific fields, like anthropology, economics, political science, and psychology, just to mention a few. Even today, his GST remains the base for the interdisciplinary study of systems in the social sciences.

A common misconception

A common misconception of von Bertalanffy's work involves misunderstanding the term "general system theory." The term translated into German is Allgemeine Systemlehre, which implies a theory of systems. This differs from the English term, which uncovers an entity termed "general system", about which he presented a theory. His overall goal was to create a general theory of systems instead of a theory of general systems, leading to the misinterpretation of his intentions. He was not attempting to present an all-encompassing theory of everything but rather a set of common principles that could be applied to different scenarios. This led to creating the term "General Systemology," coined by Manfred Drack, a theoretical biologist, and David Pouvreau, a philosopher, to address these concerns.[lxix]

Although having different beginnings, both von Bertalanffy and the cyberneticists, such as Wiener and McCulloch, share the common aim of overcoming specialization and integrating multiple scientific fields. The beginnings of first order cybernetics involved engineering with the development of theoretical models and methods that could be applied to different scientific fields with relative ease. von Bertalanffy's "systemological" project was more ideology and-based philosophy, concerning itself with the theory of knowledge, metaphysics, and axiology,[2] later expanded and added to the fields of biology and psychology.[lxx]

Second-order cybernetics, one of constructivism's main theories, indicates learners construct knowledge rather than passively absorb information. Von Bertalanffy's theory of perspectivism states the knowledge of a subject is inevitably partial and limited by the individual perspective. These two theories share common ground. Examining both fields scientifically, cybernetics begins with a mechanistic view of different types of systems, including technical devices, organisms, brains, and social patterns. Systemology first takes an organismic view of various systems, including organisms, biocenoses, psyche, and cultures.[lxxi]

---

[2] "Axiology, also called Theory Of Value, the philosophical study of goodness, or value, in the widest sense of these terms. Its significance lies (1) in the considerable expansion that it has given to the meaning of the term value and (2) in the unification that it has provided for the study of a variety of questions—economic, moral, aesthetic, and even logical—that had often been considered in relative isolation."
https://www.britannica.com/topic/axiology

Understanding all theories can be applied incorrectly, von Bertalanffy warned of the potential misinterpretations that could arise from his research. He expressed that some models may be more dangerous than others, depending on the context of their origin and reception, as some individuals can be more influenced by concepts than others. This opinion was most likely molded significantly by von Bertalanffy's experiences with Nazism. Ultimately, von Bertanaffy believed GST could become a new paradigm in thinking, with significant epistemological inference.[lxxii]

Jay W. Forrester

"The pioneering days in digital computers were exciting times. Computer development was part of the last hundred years of technological discovery. However, the major challenges facing society will not be solved by still more technology. The next hundred years will be the age of social and economic discovery. The field of system dynamics, with which I have been associated since 1956, has pioneered in understanding how organizational structures and decision-making policies interact to produce desirable and undesirable behavior in physical, biological, environmental, business, and social systems. I see this next frontier in social systems as far more exciting and important than was the technological frontier." – Jay Forrester

Jay Wright Forrester was born in the United States on July 14, 1918 in Anselmo. He was a pioneering electrical engineer and systems scientist from the US credited with inventing magnetic core memory. This technological feat is the most extensively used type of random-access computer memory that can be found in most modern digital computers. It was one of many similar devices that employed material magnetic properties to bridge the gap between vacuum tubes and semiconductors, allowing switching and amplification between the two devices. He is credited with developing the world's first computer graphics animation, a "bouncing ball" on an oscilloscope. He also contributed to developing the field of systems dynamics and oversaw the construction of the world's first general-purpose computer, which was named after him.[lxxiii]

Forrester developed an early interest in electrical systems while growing up on a cattle ranch, and he spent his spare time tinkering with batteries and telegraphs. During his senior year of high school, he delivered electricity to his family's farmhouse for the first time, using a wind-powered, 12-volt electric system he made from recycled vehicle components. He had been granted a scholarship to attend an agricultural university, but he chose to follow his passion of becoming an engineer at the University of Nebraska. He got a grasp of complex system behavior while studying theoretical dynamics in electrical engineering. In 1949, he was inducted into the Eta Kappa Nu (HKN) honors organization, which is dedicated to electrical and computer engineering, after receiving his Bachelor of Science in Electrical Engineering degree in 1939.[lxxiv]

He enrolled in graduate school at the Massachusetts Institute of Technology (MIT) as soon as he finished college and worked there until he retired in 1989. He became a research assistant at the Servomechanism Laboratory with a small salary. There, he met and worked with Gordon S. Brown, an early developer of automatic-feedback control systems and machine tool numerical control, who expanded Forrester's knowledge of the field.[lxxv]

During WWII, Brown and Forrester worked together to develop servomechanisms for regulating radar antennas and gun placements. Forrester was exposed to the first-hand transformation of research and theory into practice. Following WWII, Forrester developed and improved several electrical products thanks to servomechanisms. These mechanical tools controlled the behavior of bigger systems using feedback.[lxxvi] From 1949 to 1951, he was the head of the Massachusetts Institute of Technology's Digital Computer Laboratory, where he oversaw the development and construction of Whirlwind I, one of the first high-speed computers created for the United States Navy to manufacture an aircraft stability analyzer.[lxxvii] During his investigation, he realized early digital computers had slow and unreliable information storage systems. This was the bottleneck problem of making machines that met the dependability and performance requirements. In 1947, a three-dimensional memory system was invented, first using glow-discharge cells and, subsequently, in 1949, using toroidal random-access coincident-current magnetic storage that is still in use today. He revolutionized the industry by using magnetic cells for both storage and switching, and this approach was used until the early 1970s.[lxxviii]

The Whirlwind I computer had passed its tests successfully by 1951. The machine's original purpose was to aid in the planning and design of the United States Air Force's Semi-Automatic Ground Environment (SAGE) Air Defense System. From 1951 until 1956, Forrester was the Director of the Digital Computer Division at MIT's Lincoln Laboratory once the project was completed. During his time in this position, he oversaw developing flying simulators.[lxxix]

In 1956, Forrester chose to study management instead of tinkering with digital computers. As he immersed himself in his studies, he became interested in industrial dynamics. He pioneered the notion of systems dynamics while at the MIT Sloan School of

Management. Systems dynamics analyzes how the structure and policies of physical, social, and environmental systems influence the development, oscillation, and stability of those systems. In contrast to traditional management research and other social sciences, direct collaboration with real-world applications aided the development of systems dynamics. He created the Systems Dynamics National Model to help people comprehend economic swings. It explains the long economic wave, also known as the Kondratieff cycle, which caused the Great Depressions of the 1830s, 1890s, 1930s, and 1990s. He developed the global System Dynamics Society as a result of his study of complex systems, which has subsequently been employed in a range of sectors, such as corporate management, economic behavior, medicine, urban expansion and decay, global population, and environmental problems.[lxxx]

His early work was influenced by numerous encounters with officials at the General Electric Corporation, in which he uncovered how distribution channel systems were disrupted, leading to erroneous swings in the number of items to be ordered. This phenomenon was known as the Forrester effect, but today, people refer to it as the bullwhip effect. Because of his research, Forrester developed techniques to use feedback control to understand management and human systems, leading to the coining of the term "industrial dynamics." [lxxxi]

His discoveries prompted him to write his first book, *Industrial Dynamics*, in 1961. The book provided information about his work in systems theory, feedback control, and dynamics. The following year, he received the Academy Management Award, which recognizes persons for outstanding achievements, professional accomplishments, and distinguished service in the Department of Management at the University of California, Berkeley. His third book, *Urban Dynamics*, released in 1969, had a significant influence because it presented the results of the theory of urban interactions and emphasized the causes of previous policy failures like no other book.[lxxxii]

Forrester garnered numerous awards over his career as he made valuable contributions to the field of electrical and electronic engineering. He earned the Medal of Honor by the Institute of Electrical and Electronics Engineers (IEEE) in 1972. He received the IEEE Computer Pioneer Award in 1982 for his contributions to the world of computers. He was designated a Fellow of the Computer History Museum in 1995 "for his perfecting of core memory technology into a practical computer memory device; for fundamental contributions to early computer systems design and development." Other awards include The Franklin Institute's Howard N. Potts Medal, the George Washington University Inventor of the Year award, and the United States National Medal of Technology. In 2006, he was inducted into the Operation Research Hall of Fame that was created in his honor.[lxxxiii]

What part did he play in the evolution of systems thinking?

Forrester's most prominent contribution to the world was System Dynamics, which is best described as a computer-based approach for offering a practical and realistic examination of change processes in systems. This approach has numerous real-world applications, organizational administration, urban planning, and environmental policy being only a few. According to Forrester, the "elevator pitch" that encapsulates the benefits and understanding of system dynamics is: "System dynamics deals with how things change through time, which includes most of what most people find important." To demonstrate why our social and physical systems work the way they do, system dynamics builds on the knowledge we have about the things in our environment and employs computer simulation to show why they behave that way. When it comes to policy decisions, system dynamics demonstrate how most of our policies are the core causes of the issues we frequently assign to outside factors. It also teaches us how to find guidelines that will help us improve our situation.[lxxxiv]

Forrester wanted to apply what he had learned about system dynamics and technology to administrative difficulties. When he worked at General Electrics, he got the chance to do that. GE assigned him his first project that required analyzing variances in home appliance sales, inventories, and workers. The company was trapped in a vicious cycle, where employee workload fluctuated between absolute overworking and absolute idleness to the point of possible employment termination. These cycles appear to bear little resemblance to the steady demand for goods and services. They discovered, using Forrester's model, delays in integrating different departments inside the corporation resulted in the establishment of positive feedback loops.[lxxxv] As part of this endeavor, a compiler was created, allowing Forrester to insert the feedback equations directly into a computer, increasing managerial decision-making abilities. One of his employees oversaw creating the compiler.[lxxxvi] This compiler evolved over time into the DYNAMO (DYNAmic MOdels) programming language that was then used to develop system dynamics simulation models.

The work involved in system dynamics is broken down into four parts:
1. Information feedback systems are a relatively recent concept.
2. It is critical to understand the process through which decisions are made.
3. When dealing with complex systems, an experimental model method is used.
4. A digital computer is used to simulate mathematical models with as much precision as feasible.[lxxxvii]

Forrester considered the original foundation to be the most important in the topic because it was directly related to his work in the realm of servomechanisms. He claims, "the effect of time delays, amplification, and structure on the dynamic behavior of a system... [and] that the interaction between system components can be more important than the components themselves."[lxxxviii] This became a critical idea as systems thinking as a field of study developed. Today, we state that system elements are more replaceable than interconnections. Also, element replacement has a smaller impact on the system than changing interconnections.

After ten years of developing the subject's theoretical foundation, Forrester published *System Dynamics* in 1968. This textbook defined the field's core principles and approaches, still foundational knowledge today. According to his textbook, the structure of the system, characterized by a set of interconnected feedback loops, drives system dynamics. The interaction of these loops determines the system's behavior. It is more accurate to say a system model is created starting with the loop structure rather than with loop elements.[lxxxix]

System dynamics is effective in real-world scenarios, such as management, urban planning, and environmental change. On the other side, its influence on social issues hasn't been meaningful. To address the unresolved social concerns, Forrester and his successor, as Director of the MIT system dynamics group, John Sterman, agreed more work was needed to build the necessary approaches and models.

One of Forrester's goals, as he got closer to retirement, was to make system dynamics accessible to students at all levels of education, with particular emphasis on elementary and middle-school-aged children. According to him, if taught early, everyone may develop an understanding of the complex dynamics of nature and human systems and then utilize that knowledge to help establish a more just and sustainable society. System dynamic modelling was included in the teaching curriculum in several educational situations.

Forrester's Mental Models and System Dynamics Simulations

Forrester's 1971 study, "Counterintuitive Behavior of Social Systems", had a significant impact on the development of computerized systems models to advise social policy. Creating such models about given issues is better than a simple discourse because models allow for a more in-depth understanding of the root causes of problems and the potential repercussions of proposed solutions. This being said, mental models are a simplification of real life, thus they are imperfect and incomplete representations. Also, they can change over time and even inside a single person within a single conversation. To match the conditions of a disagreement, the human mind develops a few associations. In reaction to the topic, the model adapts.

Everyone uses mental models daily. When making decisions in their personal and professional lives, folks rely on their pre-programmed mental models to help them interpret reality as they know it. The models are built using mental representations of their personal experiences and environment. These mental models are peppered with judgments and biases. The human mind, unlike computers, is not equipped to comprehend the ramifications of a particular mental model-based solution to a complex system problem.

When discussing a particular issue, each member adopts a separate mental model to interpret the circumstances. When people's basic assumptions are so different, yet so seldomly brought into the conversational light, it's not surprising that getting to a consensus regarding problem-solving takes a long time. Even when an agreement is reached, it may create legislation and programs that either fail to fix the problem or create unintended consequences.[xc] In other words, the human mind has its limitations, having the tendency to jump to the wrong conclusions – either individually or as a group – led by emotional reasoning. When the human mind creates a computer model to replicate its assumptions, the system often acts differently than anticipated. This is because "there are internal contradictions in mental models between assumed structure and assumed future consequences."[xci]

System dynamic stimulation models incorporate all assumptions and how they interact and impact the system. These computer-based simulations can minimize ambiguity, expose hidden assumptions, and make space for debates. Computer models surpass mental models in terms of predicting future dynamic effects caused by the interacting assumptions.

Instead of just another theory, the Forrester Group (did you read Forrest Gump first?) built an entirely new way of thinking. This way of thinking has made a substantial difference in understanding and modelling real-world problems despite its concentration on computer-based modelling.[xcii]

Peter M. Senge

Peter Michael Senge was born in Stanford, California in 1947. He is a Senior Lecturer at the Massachusetts Institute of Technology and a renowned systems scientist. He is the Founder and Chair of the Society for Organizational Learning (SoL), a global network of companies, academics, and consultants dedicated to the "interdependent development of people and their institutions."[xciii]. He is also a co-founder and board member of the Academy for Systems, a non-profit that assists leaders in developing their capacity to lead in complex social systems that promote biological, social, and economic well-being. This organization is dedicated to developing tools, processes, and approaches for awareness-based systems thinking.[xciv]

Stanford University awarded Senge a Bachelor of Science in Aerospace Engineering. During his time at Stanford, he also studied philosophy. As a student, he became interested in population growth, which led him to explore the world's current key issues, such as overcrowding, hunger, and the environment. Later, he went on to MIT, where he received a Master of Science in social systems modeling in 1972, followed by a PhD in Management from the MIT Sloan School of Management in 1978.[xcv]

After receiving his degree, Senge worked as an engineer in training for John H. Hopkins. Michael Peters and Robert Fritz, both composers and researchers, supported him, and he closely followed their work. He obtained practical experience with firms such as Ford, Chrysler, Shell, AT&T, Hanover Insurance, and Harley-Davidson in the 1970s and 1980s, basing his works on pioneering work with the five disciplines about which he later wrote his groundbreaking book. He continued his studies at MIT while working as an engineer.[xcvi]

Senge's first revolutionary book, *The Fifth Discipline: The Art and Practice of the Learning Organization*, was published in 1990. As a result, he became well-known in organizational development. He developed the phrase "learning organization," which refers to a business that encourages its personnel to learn and evolve. Learning organizations may emerge because of the demands that modern businesses face, allowing them to remain competitive in the marketplace. According to Senge, only businesses that can adjust quickly and efficiently will thrive in their field or market. Two requirements must be met at all times for a corporation to be considered a learning organization. The first is the ability to create the organization to match the expected results. The second is the skill to see when the organization's initial course deviates from the planned results and then rectify this deviation.[xcvii] A learning organization is one in which "people continually expand their capacity to create the results they truly desire; where new and expansive patterns of thinking are nurtured; where collective aspiration is set free; and where people are continually learning how to learn together."[xcviii]

His debut book, *The Fifth Discipline*, was named by the Harvard Business Review "one of the important management books of the last 75 years," and the Financial Times named it "one of the five "most essential" management books in 1997. Senge was also named "Strategist of the Century" by the *Journal of Business Strategy*, being one of few who "had the greatest impact on the way we conduct business today."[xcix] He was recently named to China's "1000 Talents" initiative, which aims to help China become a leader in systemic change to benefit China and the rest of the world.[c]

The Fifth Discipline gives ideas and methods for fostering ambition, inspiring reflective dialogue, and grasping complexity to build a learning organization. Too many businesses are considering the CEO a savior, who can rescue the company and push staff to change. As a result, grandiose goals are initiated but never realized. The effort is eventually overcome by resistance. Senge contends that cultivating reflection and inquiry skills is crucial in allowing for discussion of real-world situations. He cites four challenges an organization needs to overcome when adopting changes:

1. There must be a compelling case for change.
2. There must be a window of opportunity for change.
3. Assistance is essential during the transformation process.
4. As the obvious barriers to change are removed, it is critical that a new issue, perhaps not considered important or even noted, does not emerge as a major block.[ci]

The Fifth Discipline Fieldbook was co-authored by Senge and published in 1994. It was a response to reader demands for further information on tools, processes, and practical examples in developing improved learning capacity within organizations. His third book, *The Dance of Change*, was published in 1999. This book talks about contemporary challenges companies with years-long learning development face and the techniques designed by leaders to deal with difficulties the sustained learning process presents. Senge has also published several books and essays on systems thinking in management in academic journals and business magazines.[cii]

Senge and systems thinking

Senge was a big supporter of systems thinking, which he has called the "cornerstone" of learning companies. Systems thinking is the connecting force of his five disciplines. Systems thinking focuses on how elements interact with other elements, trying to achieve a function or purpose within the system. Instead of focusing only on people within an organization, learning organizations use a broader approach to studying a wider range of interactions both within and between companies.

Senge's main purpose was to figure out how we can all work together to coexist. He gives workshops and seminars around the world, sharing his ideas with inspiring global learning groups. He discusses how to integrate abstract principles from systems theory into practical tools for better understanding economic and organizational transformation. His study focuses on decentralizing leadership in organizations to improve everyone's ability to work toward common goals. It also promotes shared knowledge of complex concerns and shared leadership for healthy systems, such as large projects focusing on global food systems, climate change, and the future of education. His organizational learning theories have helped businesses, educational institutions, health care facilities, and governments.[ciii]

Senge's work as a management professor and consultant brought systems thinking to the attention of a wide audience. Senge identifies himself as an "idealistic pragmatist" with the idealistic goal of finding out "how can we build the capacity to address the common systemic issues that are taking our society in directions no one really wants to go."[civ]

Senge's first book highlights seven critical "learning disabilities" that must be addressed to improve:

1. "I am my position": Over time, people in organizations develop a strong attachment to their job; they become what they do. As a result, they have a myopic and non-systemic view of the organization, losing sight of how their actions affect the system.
2. The "The 'enemy is out there' syndrome" is a byproduct of the "I am my position" worldview. When one defines themselves by their job, when things go wrong, they assume someone (else) "screwed up." This leaves people with a restricted sense of self-

identification. It's not hard to regard those around us as opponents.

3. "The Illusion of Taking Charge" - "proactiveness" often implies "I'm going to get more active in fighting those enemies out there." Seeing how one's actions affect the system is what true proactiveness entails.

4. "Fixation on events" – People tend to see life as a series of events, each with a single obvious explanation. It's what we learn. The irony is that the biggest risks to human survival – in both organizations and civilizations – are caused by long, continuous processes rather than by startling catastrophes.

5. "The Parable of the Boiled Frog": Humans are good at responding to immediate threats but are terrible at detecting gradual threats, such as the frog that will sit in a kettle of water and slowly boil to death. (This parable proved to be untrue – the frog does leap out once the water reaches a critical temperature. In reality, the frog is more harmed when being dropped in boiling water as it gets severely burnt despite jumping out.[cv])

6. "The Delusion of Learning from Experience" – We learn best by experience, through trial and error, yet we never see the consequences of our most important decisions. The most

critical decisions made in organizations have long-term consequences for the entire system.
7. "The Myth of the Management Team" - Most teams operate at the level of the group's lowest skilled person. As a result, "skilled incompetence" emerges - groups of people who are competent at learning.[cvi]

Senge described the five disciplines that may be used to assist people with learning disabilities:
1. Personal growth and development—personal mastery.
2. Mental models are representations that help people understand their surroundings and optimize solutions to their problems.
3. Personal visions inspire the development of shared visions.
4. Team learning has three critical components.
    a. The requirement to think critically about challenging issues
    b. The requirement for innovative and well-organized action
    c. Through team members' responsibilities on other teams, learning is always fostered.

5. In systems thinking, the other disciplines come together to form a cohesive body of theory and practice.[cvii]

The investigation of mental models is based on the results of management specialists Chris Argyris and Donald Schön in the fields of action science and reflective practice. Team learning is based on Argyris' work on organizational defense mechanisms and physicist David Bohn's study on how to stimulate group argument as a form of learning. Senge's work at Innovation Associates gave birth to the disciplines of personal mastery and shared visioning. Finally, systems thinking is a modified version of system dynamics in which systems are diagrammed and a series of interrelated feedback loops are illustrated. These diagrams are frequently classified as system archetypes. According to systems thinking, "the discipline that integrates the disciplines, fusing them together into a coherent body of theory and practice." Senge's usage of the word "systems thinking" was noteworthy because it was not commonly used at the time. Senge only used it to define his modified form of system dynamics, even though it hints to a broader variety of system approaches.[cviii]

Senge's use of system dynamics has caused debate, particularly because he presents system dynamics without using simulation models. "The danger comes from encouraging people to believe that systems thinking is the whole story... It's only from the actual simulations that inconsistencies within our mental models are revealed." Forrester describes Senge's use of the area.. "Systems thinking can be a first step toward a dynamic understanding of complex problems, but it is far from sufficient."[cix]

Despite its criticism, Senge's work has spurred interest in the subject, bringing systems thinking to mainstream audiences. He has altered people's perceptions about organizational learning and the role of systems thinking in it.

Four examples of system archetypes.[cx]

1) "Today's Problems Come from Yesterday's 'Solutions' " – When we try to identify the root cause of a problem, we often need to look to prior solutions. For example, law enforcement officers tackling drug issues is an example of this. Shutting down a drug business on A street only shifts it to B street. Catching drug smuggling only exasperates the situation, increasing drug price and drug-related crimes by desperate addicts. The solution – catching drug dealers and diminishing the amount of drugs on the streets

– only shifted the problem into a new dimension, and now new tools are needed to handle the new problem.

2) "The Harder You Push, the Harder the System Pushes Back" – It's a popular belief that, to achieve success in life, you have to work harder. No success? Well, you haven't worked hard enough. While this adage stands true to a certain extent, there comes a time when it backfires – the harder you work, the more work you'll have as people will trust you're the go-to person when grinding is needed. Others will profit on your back. This phenomenon is known as "compensating feedback" in systems thinking. The more energy one invests in trying to fix the problem, the more energy seems to be demanded.

3) "Behavior Grows Better Before It Grows Worse" – Some interventions seem promising in the short-term only to turn sour in the long-run. One can give a fish to a hungry person today and feel they sacrificed at the altar of good karma, but this act won't help the situation of such a person tomorrow. Compensating feedback involves a delay between short-term gains and long-term losses. Lags within a system are difficult to

detect. The first intervention that seemingly fixes the issue only treated a symptom. It's only a matter of time for the issue to reappear or a new, bigger problem to emerge.

4) "The Easy Way Out Usually Leads Back In" – It's natural to apply known solutions to difficulties. It's quick, easy, and may have worked in the past. However, this way of thinking is nonsystemic thinking, often known as the "what we need is a bigger hammer" syndrome.[cxi] While the familiar remedy is repeatedly attempted, the initial issue continues or worsens.

Professor Russell L. Ackoff

Russell L. Ackoff was born in the city of Philadelphia, Pennsylvania on February 12, 1919. He received a bachelor's degree in architecture from the University of Pennsylvania in 1941, and after graduation, he worked as an assistant professor of philosophy at Penn for a year. He was a member of the United States Army stationed in the Philippine Islands from 1942 until 1946. Before attending Officer Candidate School, he served in the United States Army's Fourth Armored Division. After returning home, he enrolled at the University of Pennsylvania, where he earned a PhD in philosophy of science in 1947. Professor Ackoff was an Assistant Professor of Philosophy and Mathematics at Wayne State University from 1947 until 1951. Ackoff has received multiple honorary doctorates from 1967 until the present.[cxii]

Ackoff and a group of colleagues were invited to join the Case Institute of Technology School of Engineering in Cleveland, which is now part of Case Western Reserve University, in 1951. They worked together to build one of the world's first Operations

Research Departments, where he was an associate professor and later a professor of Operations Research. In honor of his achievements, Ackoff is now known as the "Father of Operations Research." Ackoff was one of the founders of the Operations Research Society of America (ORSA) and served as its fifth president from 1956 until 1975. In 1987, he was elected president of the International Society for Systems Sciences (ISSS). He also collaborated with colleagues Churchman and Leonard Arnoff on the first textbook in the discipline, *Introduction to Operations Research*, which was published in 1957 and is still in print today.[cxiii]

Ackoff was a professor of systems sciences and management sciences at Wharton School at Penn for most of his stay there. His work was not limited to a single style, and he was concerned with ethical and social issues throughout his career. Because the mathematical technique employed in Operations Research became more precisely defined, Ackoff coined the term "Social Systems Science" to designate the study of social systems. While distancing himself from Operations Research in the 1970s, he became one of the most prominent critics of "technique-dominated Operations Research" and advocated for more participatory approaches to problem solving. Wharton's Social Systems Sciences Program was well-known for combining theory and practice, breaking down academic boundaries, and encouraging students to think and act independently. His disillusionment with Operations Research guided Ackoff to systems thinking.[cxiv]

Professor Russell L. Ackoff was well-versed in many areas of study, including architecture, philosophy, city planning, behavior science, organizational operations, and systems thinking, to name a few. He spent the most of his working life attempting to overcome severe social and organizational problems through collaboration with all parties concerned.[cxv]

Ackoff was the first PhD student of C. West Churchmans, a philosopher and systems scientist from the United States. Both Churchman and Ackoff championed "experimentalism," a philosophical approach described by philosopher Edgar A. Singer, Jr. that contends t the only way to uncover the truth is through experiments and empiricism. He put experimentalism into action by founding "institutes of experimental method," dedicated to applying philosophical assumptions about man's nature to the construction and reinforcement of social structures.

When Ackoff retired as Anheuser Busch Professor Emeritus of Management Science at the Wharton School of Business in 1986, he launched the

Institute for Interactive Management (INTERACT). This allowed him to continue his consulting and academic work, including over 350 firms and 75 government organizations in the US and internationally, as well as his research. In addition, he was a visiting professor of marketing at Washington University in Saint Louis from 1989 to 1995. According to the *Financial Times*, he was placed 26th on a recent list of the world's most prominent business thinkers for his management abilities. [cxvi]

In 2007, the Russell L. Ackoff Systems Thinking Library and Archive opened its gates in the Organizational Dynamics program of the School of Arts and Sciences. In the same year, the Ackoff Department was launched at Tomsk University in Tomsk, Russia. A year later, in 2008, the Russell L. Ackoff Systems Thinking Library and Archive become the collecting hub of approximately three dozen of Ackoff's writings and his private papers and a personal library of about 3000 books on systems, design, philosophy, and social science. In 2008, the New Bulgarian University in Sofia, Bulgaria developed an additional Ackoff Program, and in 2009, the Da Vinci Institute in Cape Town, South Africa established the Ackoff Center for Design Thinking. Ackoff continued to teach throughout his life.[cxvii]

During the Clinton and Bush administrations, Ackoff's concepts were implemented in the White House Communications Agency and the White House Military Office in collaboration with Dr. J. Gerald Suarez, a Professor of Practice in Systems Thinking and Design at the University of California, Berkeley. Attempts were made to usher the White House into the age of systems thinking.[cxviii]

How does he fit within the subject of systems thinking?

Ackoff was a key figure in the field of systems thinking as a significant proponent of using systems methodologies to account for the complexity of interrelated problems rather than just delivering technical answers. Aside from his well-known status as a theorist, his exceptional work as a consultant gave him the term "guru," which he despised, preferring to call himself an educator. "A consultant goes in with a solution," he added. "He tries to impose it on a situation. An educator tries to train the people responsible for the work to work it out for themselves. We don't pretend to know the way to get the answer."[cxix]

His colleagues recognized him as an excellent communicator, verbally and in writing. He regularly employed anecdotes, aphorisms, and well-argued notions to illustrate his views in his works. This collection of short publications resulted in the publication of *Ackoff's Fables* and *Systems Thinking for Curious Managers*.

Ackoff co-authored the paper "On Purposeful Systems: An Interdisciplinary Analysis of Individual and Social Behavior as a System of Purposeful Events" with Australian psychologist Frederick Edmund Emery in 1975. It was a seminal work on purposeful systems. The emphasis of this article was on the relationship between systems thinking and human behavior. "Individual systems are purposive and knowledge and understanding of their aims can only be gained by taking into account the mechanisms of social, cultural, and psychological systems."[cxx] Any human-created system can be categorized as a purposeful system if the "members are also purposeful individuals who intentionally and collectively formulate objectives and are parts of larger purposeful systems."[cxxi] The fast changes in these systems may have led to the end of the "Machine Age" that dominated since the Industrial Revolution and the beginning of the "Systems Age." Reductionism (everything can be reduced to parts that can't be further divided) and mechanism (cause-effect relationships) – the dominant ideas of the industrial age - have given way to expansionism (the belief that all objects and events, and all experiences of them, are parts of larger wholes) and teleology (an explanation for something that serves as a function of its end, purpose, or goal, as opposed to an explanation for something that serves as a function of its cause).[cxxii] This period, according to Ackoff, is distinguished by the emergence of a new conceptual framework.

Ackoff argued humanity leaving the Machine Age and entering the Systems Age meant a shift from analytical thinking and reductionism to systems thinking and expansionism. Due to this shift, Ackoff coined a new term in systems thinking, "mess." "We have also come to realize that no problem ever exists in complete isolation. Every problem interacts with every other problem and is therefore part of a set of interrelated problems, a system of problems. Furthermore, solutions to most problems produce other problems; for example, buying a car may solve a transportation problem but it may also create a need for a garage, a financial problem, a maintenance problem, and conflict among family members for its use."[cxxiii]

Breaking down a problem – or mess - into its smallest elements may make things worse rather than help to fix it. Ackoff used a range of instances from urban life to demonstrate his point. Healthcare, education, transportation, and crime are just a few examples where this dynamic plays out. Rather than attempting to fix these issues one at a time, he suggested a holistic "mess management" strategy be used. Planning, notably "interactive planning," which entails a participatory and systematic approach, can help him prevent problems in the first place, he claims. This technique, along with the concept of participatory management, formed the basis of Ackoff's systems thinking work in the 1970s and beyond.[cxxiv]

Ackoff's 'idealized redesign' of a system means, throughout this process, every aspect of the system and the input of all persons involved need to be considered. He outlined five stages of interactive planning:

1. Recognizing the obstacles and opportunities involved in the "mess."

2. What is the "ends" of planning? What kind of future do we envision?

3. What are the "means" of planning? This phrase is the process of devising strategies to achieve the desired future.

4. Resource planning - what resources are needed to achieve the desired future and how to obtain those resources.

5. Iterative implementation and control design — the mechanics of putting the changes in place and ensuring they are carried out effectively.[cxxv]

6.

Ackoff's other contributory ideas to systems thinking include:

- "Improving the performance of the parts of a system taken separately will not necessarily improve the performance of the whole; in fact, it may harm the whole.

- Problems are not disciplinary but are holistic.

- The best thing that can be done to a problem is not to solve it but to dissolve it.

- The healthcare system of the United States is not a healthcare system; it is a sickness and disability-care system.

- The educational system is not dedicated to produce learning by students, but teaching by teachers—and teaching is a major obstruction to learning.

- The principal function of most corporations is not to maximize shareholder value, but to maximize the standard of living and quality of work life of those who manage the corporation."cxxvi

Peter B. Checkland

Peter Checkland was born in Birmingham, United Kingdom on December 18, 1930. One of his greatest contributions to systems thinking is the development of the soft systems methodology (SSM), based on a holistic approach to thinking and practicing in soft systems environments. According to the philosophy, real-world problems can be solved by employing solutions that are both systemically desirable and culturally acceptable.[cxxvii]

He was educated at George Dixon's Grammar School and, after national service, at St. John's College, Oxford, where he earned a Master of Arts degree in 1954 with first-class honors in chemistry. His first position was as an assistant professor of systems engineering at Lancaster University. The department offered one-year master's degrees in which students applied systems concepts in a real-world setting. During the next 30 years, he collaborated with colleagues and master's students on an "action research" program, a concept and method of study common in the social sciences. It promotes revolutionary change by allowing individuals to act and observe, with critical reflection acting as a bridge between the two. [cxxviii]

Checkland wrote more than a dozen articles and four books, the first of which was published in 1981, titled *Systems Thinking, Systems Practice*. All these document his contributions to the evolution of SSM. His third book, *Information Systems and Information Systems*, was co-authored with Sue Holwell, a British novelist, and published in 1998. He also served on the editorial boards of several journals, including the *European Journal of Information Systems*, the *International Journal of Information Management*, the *International Journal of General Systems*, the *International Journal of General Systems Practice*, and the *Systems Research Journal*. Checkland was elected president of the Society for General Systems Research (now known as the International Society for the Systems Sciences) when it was founded in 1986.[cxxix]

As a result of his work, Checkland has received several honors, four honorary doctorates from City University; the Beale Medal of the Operational Research Society in 2007 (for his significant contribution to the philosophy, theory, and practice of operational research); the Gold Medal of the UK Systems Society; and the Pioneer Award of the International Systems Society. Aside from that, he was one of the Omega Alpha Society's first four Fellows.[cxxx]

Checkland's contribution to systems thinking.

Checkland has had a significant impact on systems thinking, particularly in management and information technology. His most significant contributions include the development of SSM and the introduction of other critical conceptual advances, such as the distinction between "hard" and "soft" systems thinking and his promotion of the soft approach.

Soft Systems Methodology, SSM, was a new technique for dealing with issues where repeated interventions prove useless to solve the problem. SSM is a type of systems thinking methodology developed to learn about real-world scenarios. It is now widely used, being at the heart of the paradigm shift from "hard" systems thinking and toward "soft" systems thinking. The need for a soft system method arose because of "hard" systems engineering's inability to address the difficulties of mess management effectively. Hard systems imply systems are part of the world. Soft systems assume systems are part of our understanding of the world. Checkland's work resulted in the development of "soft" operations research, which incorporated techniques such as optimization, mathematical programming, and simulation into the overall operations research map. "Rather than SSM models attempting to map the real world – impossible because there are multiple candidates for what counts as the real world in complex situations – the models are devices for learning about the real world. In short, SSM becomes a process of inquiry, a learning system."[cxxxi]

Checkland makes a clear distinction between hard systems thinking and soft systems thinking throughout his first book, *Systems Thinking, Systems Practice*. "Hard systems thinking" refers to methodologies such as systems engineering and operational research that simply accept the issue or demand as it is and do not modify it. The true nature of the problem is revealed during the process of system analysis, which is a component of soft systems thinking.

SSM acknowledges the "process of inquiry," originally established as a learning system. In 1985,

Checkland defined inquiry as consisting of three parts: an intellectual framework (F), a methodology for applying that framework (M), and an application area (A) in which the methodology is used.[cxxxii]

Our daily lives are affected by complex, interacting events and thoughts that can lead to systemic problems. We can approach these difficulties in a systematic manner by employing SSM. It can assist us in organizing our thoughts so we improve our situation. Systemic problems are difficult to understand because they are never static and involve multiple interacting conceptions of "reality." Different people have different assumptions about the world and, as a result, have different views. Another feature of systemic problems are conflicting worldviews, and frequently witnessed attempts at meaningful acts offer guidance in dealing with adversity. Checkland's original SSM paradigm included seven-steps he referred to as the "learning cycle." He highlights there are multiple ways to practice SSM, and his initial seven steps is just one of them. These steps include the following:

1. Understand the problem in an unstructured manner.
2. Express the process and structure of the problem in a formal way. Apply rich visuals to

facilitate understanding. (i.e., actual drawings that show the features of a problem).
3. Create root definitions and identify relevant systems. This point can be further analyzed using the mnemonic CATWOE:
   a. Customers - the system's beneficiaries.
   b. Actors within the system – who execute the significant actions in the system.
   c. Transformation of given inputs into specified outputs.
   d. Worldview - a perspective, framework, or vision that provides meaning to a fundamental concept. It is also known as a point of view, framework, or vision.
   e. Ownership -a person or entity who has a primary interest in the system and the authority over the existence of the system.
   f. Environmental constraints - characteristics of the system's surroundings that are given.
4. Build conceptual models for each root definition to represent the ideal state of the system.
5. To compare and contrast the conceptual models with the initial depiction of the problem.

6. Based on this comparison, identify culturally viable and systemically optimal changes to try to reach the ideal state of the system.
7. Take action.[cxxxiii][cxxxiv]

Checkland continued to research and develop SSM and related concepts throughout the 1980s and 1990s. It enabled him to express the distinction between hard and soft systems thinking in terms of whether the systems were a part of the world (hard) or a part of our understanding of the world (soft), later describing it as a "shift of systemicity from the world to the process of inquiry into the world." This shift is significant because it reflects the philosophical school's move from functionalism (or positivism). It is part of Checkland's reorientation of the field of systems thinking toward the philosophical school of interpretivism that examines social work from multiple perspectives rather than the functionalist approach, which suggests there is only one "real" view of the world.[cxxxv]

Ervin László

Ervin László was born in the Hungarian capital of Budapest on May 12, 1932. He is a Hungarian scientific philosopher, systems theorist, integral theorist, classical pianist, and advocate for quantum consciousness theory. His father was a cobbler. His mother was a pianist, who taught him to play the piano. László made his debut with the Budapest Symphony Orchestra when he was nine years old. He was granted permission to leave Hungary after winning the Grand Prize in an international music competition in Geneva, and he embarked on a worldwide performing career that took him first to Europe and then to the United States. He immigrated to the United States after World War II. László received the state doctorate in philosophy and human sciences from the Sorbonne University in Paris, the highest honor in the fields of philosophy and humanities at the time. In addition, he was awarded the prestigious Artist Diploma from the Franz Liszt Academy in Budapest.[cxxxvi]

He transitioned from music to a career as a scientist and humanist, giving presentations at universities all over the United States, including Yale and Princeton. He worked on a model of future world order while at Princeton, and he was later asked to submit a report to the Club of Rome, of which he was a member. He worked on global initiatives at the United Nations Institute for Training and Research, established at the request of the Secretary-General in the late 1970s and early 1980s.

He discovered the existence of the Akashic Field while researching. The Akashic Field is the concept of an interconnected cosmic field at the roots of reality that contains information preserved by the world's sages. All universal events that have ever occurred in the past, present, or future and all thoughts and words that have ever occurred among them are included in the database.[cxxxvii] *Science and the Akashic Field: An Integral Theory of Everything*, published in 2004, expresses his belief that the "quantum vacuum" is the underlying energy and information-carrying field that orients not only the current cosmos, but all universes past and present. The author says it is because of such an informational field that the universe appears to be fine-tuned, capable of producing galaxies and conscious lifeforms, and evolution does not happen at random.[cxxxviii]

In 1984, László and some of his colleagues formed the initially covert General Evolutionary Research Group. The goal of this research group was to see if chaos theory could be used to find a new general theory of evolution that could lead to a more just and equitable world.[cxxxix]

Because of his positive experience with the Club of Rome, László founded the Club of Budapest in 1993. The mission of the Club of Budapest is "facilitating and providing direction to a „global shift" toward a more peaceful, equitable, and sustainable world." He is also the president of the László Institute for New Paradigm Research and has served as president of the International Society for Systems Sciences. László has received honorary doctorates from several countries, including the United States, Canada, Finland, and Hungary. He was given various honors and awards, including the Goi Award, the Japan Peace Prize in 2001, the Assisi Mandir of Peace Prize in 2006, and nominations for the Nobel Peace Prize in 2004 and 2005. In addition, he received the Luxemburg Peace Prize in 2017. László was named one of the "100 Most Spiritually Influential Living People in the World" by *Watkins Mind and Body Spirit* magazine in 2019.[cxl]

Throughout his career, he held a few guest professorships at universities in Europe and Asia. He was also a professor in philosophy, systems sciences, and future sciences at the Universities of Houston,

Portland State, and Indiana, Northwestern University, and the State University of New York. He has held the position of Director-Adviser for the United Nations Educational, Scientific, and Cultural Organization (UNESCO), the General's Ambassador for the International Delphic Council, and was a member of the International Academies of Science, Arts, and Sciences, and Philosophy. László is the author or co-author of over 101 publications in approximately 23 languages and the editor of another 30 volumes. Aside from that, he has published several papers and pieces in both scholarly journals and popular periodicals.[cxli]

How does he fit into the field of systems thinking?

László was the first person to make a formal introduction to the field of systems philosophy, publishing *Introduction to Systems Philosophy: Toward a New Paradigm of Contemporary Thought* in 1972. The goal of this discipline was described as a "reorientation of thought and world view resulting from the introduction of 'systems' as a new scientific paradigm." László defines system philosophy as using a systems approach to model the nature of reality and address global challenges. László developed his ideas on the subject independently of von Bertalanffy's GST work, but the two met before László published his book and chose the name systems philosophy.[cxlii]

There are four main domains in systems philosophy:

1: "Systems ontology:" this domain seeks to discover what one means by "system" and "how systems are realized at various levels of the world of observation"[cxliii]

2: "Systems paradigms:" this area wants to create a worldview that "takes [humankind] as one species of concrete and actual system, embedded in encompassing natural hierarchies of likewise concrete and actual physical, biological, and social systems";[cxliv]

3: "Systems axiology:" this domain aims to develop system models on symbols, values, social cultures, and humanistic concerns. This has to be embedded in "cosmic hierarchies."[cxlv]

4: "Applied systems philosophy:" this is the practical and action-oriented area that seeks to solve the problems identified in the three areas above.

László meant to utilize a systems view to describe reality through his "systems philosophy" in an attempt to fix the critical problems of humanity. Due to general system theory's capturing of recurring patterns across various fields of study, László deduced that the world has an underlying unity.

Therefore, all natural domains are "expressions, arrangements, or projections" of an intelligently ordered reality. If this unity could be understood, we would have much better solutions to problems in society and could answer challenging philosophical questions.[cxlvi]

References

A. (2019, May 18). *What is the Akashic Field? by CJ Martes*. TOWARDS LIFE-KNOWLEDGE. Retrieved May 11, 2022, from https://bsahely.com/2016/11/13/what-is-the-akashic-field-by-cj-martes/

About Russell Ackoff. (n.d.). Triarchy Press. Retrieved May 11, 2022, from https://www.triarchypress.net/ackoff.html

Abraham, T. (2016, October). *Rebel Genius*. The MIT Press. Retrieved April 30, 2022, from https://mitpress.mit.edu/books/rebel-genius

Ackoff, R. (2015, November 19). *A Lifetime of Systems Thinking*. The Systems Thinker. Retrieved May 11, 2022, from https://thesystemsthinker.com/a-lifetime-of-systems-thinking/

Ackoff, R. L. (1974). *Redesigning the future: a systems approach to societal problems* (Ex-library ed.). Wiley.

ASQ: About: W. Edwards Deming | ASQ. (n.d.). ASQ. Retrieved May 2, 2022, from https://asq.org/about-asq/honorary-members/deming

B. (2021, March 29). *Margaret Mead*. Biography. Retrieved May 2, 2022, from https://www.biography.com/scholar/margaret-mead

Boulding, R. J. (2018). *Elise Boulding: A Pioneer in Peace Research, Peacemaking, Feminism, Future Studies and the Family: From a Quaker Perspective (Pioneers in Arts, Humanities, Science, Engineering, Practice, 6)* (Softcover reprint of the original 1st ed. 2017 ed.). Springer.

British Library. (n.d.). The British Library. Retrieved May 2, 2022, from https://www.bl.uk/people/w-edwards-deming

Checkland, P. (1999). *Systems Thinking, Systems Practice: Includes a 30-Year Retrospective* (1st ed.). Wiley.

Clyburn, G. (2018, January 8). *The Fifth Discipline*. Carnegie Foundation for the Advancement of Teaching. Retrieved May 10, 2022, from https://www.carnegiefoundation.org/blog/the-fifth-discipline/

Deming Collaboration. (2018, August 20). *Deming's 14 Points*. Retrieved May 2, 2022, from https://demingcollaboration.com/w-edwards-deming/demings-14-points/

Edwards Deming. (n.d.). Global Association for Systems Thinking. Retrieved May 2, 2022, from https://545550393541748825.weebly.com/edwards-deming.html

Ervin Laszlo. (n.d.-a). *The Club of Budapest*. Retrieved May 11, 2022, from https://www.clubofbudapest.com/ervin-laszlo

Ervin Laszlo. (n.d.-b). *Ervin Laszlo*. Retrieved May 11, 2022, from https://ervinlaszlobooks.com/author/ervin-laszlo

Ervin Laszlo. (2021, October 22). *The Laszlo Institute*. Retrieved May 11, 2022, from https://thelaszloinstitute.com/about/ervin-laszlo/

Forrester, J. W. (2013). *Industrial Dynamics*. Martino Fine Books.

G. (n.d.-a). *Ludwig von Bertalanffy*. Environment and Ecology. Retrieved May 3, 2022, from http://environment-ecology.com/biographies/395-ludwig-von-bertalanffy.html

General Education Research Group. (n.d.). *The Darwin Project*. Retrieved May 11, 2022, from https://www.thedarwinproject.com/gerg/gerg.html

GoodTherapy Editor Team. (2018, May 3). *Gregory Bateson (1904–1980)*. The Good Therapy. Retrieved May 2, 2022, from https://www.goodtherapy.org/famous-psychologists/gregory-bateson.html

Grant, A. (2021). *Think Again: The Power of Knowing What You Don't Know*. Viking.

Gregory Bateson - New World Encyclopedia. (2008). New World Encyclopedia. Retrieved May 2, 2022, from https://www.newworldencyclopedia.org/entry/Gregory_Bateson

History.com Editors. (2019, October 28). *Margaret Mead*. HISTORY. Retrieved May 2, 2022, from https://www.history.com/topics/womens-history/margaret-mead

I. (n.d.-b). *Ackoff, Russell L*. INFORMS. Retrieved May 11, 2022, from https://www.informs.org/Explore/History-of-O.R.-Excellence/Biographical-Profiles/Ackoff-Russell-L

I. (n.d.-c). *Forrester, Jay W*. INFORMS. Retrieved May 9, 2022, from https://www.informs.org/Explore/History-of-O.R.-Excellence/Biographical-Profiles/Forrester-Jay-W

infed.org & infed.org. (2013, February 16). *Peter Senge and the learning organization – infed.org:* Infed. Retrieved May 10, 2022, from https://infed.org/mobi/peter-senge-and-the-learning-organization/

International Bateson Institute. (2018, November 14). *Gregory Bateson*. The International Bateson Institute. Retrieved May 2, 2022, from https://batesoninstitute.org/gregory-bateson/

Jaques, M. L. (1999). *Transformation and Redesign at the White House Communications Agency.* Quality Management Journal, Volume 6, Issue 3. Retrieved May 11, 2022, from https://curiouscat.net/pdfs/management/marchqmj99.pdf

Kaiser, A. (2018, May 14). *Learning from the future meets Bateson's levels of learning | Emerald Insight.* Emerald. Retrieved May 2, 2022, from https://www.emerald.com/insight/content/doi/10.1108/TLO-06-2017-0065/full/html

Kang, P. (2021, January 24). *The Seven Learning Disabilities from The Fifth Discipline.* Peter Kang. Retrieved May 10, 2022, from https://www.peterkang.com/the-seven-learning-disabilities-from-the-fifth-discipline/

Kirby, M., & Rosenhead, J. (2005). IFORS Operational Research Hall of Fame: Russell L. Ackoff. *IFORS Operational Research Hall of Fame: Russell L. Ackoff, 12*(Intl. Trans. in Op. Res), 129–134.

Laszlo, E. (1972). *Introduction to Systems Philosophy: Toward a New Paradigm of Contemporary Thought* (1st ed.). Gordon and Breach.

Laszlo, E. (2004). *Science and the Akashic Field: An Integral Theory of Everything* (Original ed.). Inner Traditions.

Lee, H. L. (1997, April 15). *The Bullwhip Effect in Supply Chains*. MIT Sloan Management Review. Retrieved May 9, 2022, from https://sloanreview.mit.edu/article/the-bullwhip-effect-in-supply-chains/

Lee, J. A. N. (2016, November 16). *Computer Pioneers - Jay Wright Forrester*. IEEE Computer Society. Retrieved May 9, 2022, from https://history.computer.org/pioneers/forrester.html

Lineausson, G. (2006). *School of Technology and Society DISSERTATION Literature Review on System Dynamics and Simulation*. University of Skovde. https://www.diva-portal.org/smash/get/diva2:1046998/FULLTEXT01.pdf

Ludwig Von Bertalanffy - Work - Open Systems | Technology Trends. (n.d.). Primidi. Retrieved May 3, 2022, from https://www.primidi.com/ludwig_von_bertalanffy/work/open_systems

Marsalli, M. (2007). *McCulloch-Pitts Neurons*. The Mind Project. Retrieved April 30, 2022, from https://mind.ilstu.edu/curriculum/mcp_neurons/index.html

McGriff - Knowledge Base - Systems Theory. (2001). Portland State University. Retrieved May 11, 2022, from https://web.archive.org/web/20070626112905/http://www.personal.psu.edu/sjm256/portfolio/kbase/Systems%26Change/systems.html

Parks, T. (2018, December 15). *Gregory Bateson changed the way we think about changing ourselves | Aeon Essays*. Aeon. Retrieved May 2, 2022, from https://aeon.co/essays/gregory-bateson-changed-the-way-we-think-about-changing-ourselves

Peter Checkland. (n.d.). Lancaster University. Retrieved May 11, 2022, from https://www.lancaster.ac.uk/lums/people/peter-checkland

Peter Checkland. (2010). Academic Dictionaries and Encyclopedias. Retrieved May 11, 2022, from https://en-academic.com/dic.nsf/enwiki/251277

Peter Checkland. (2015, June 18). Vanguard Consulting Ltd. Retrieved May 11, 2022, from https://beyondcommandandcontrol.com/library/whos-who-system-thinkers/peter-checkland/

Peter M. Senge. (2019, April 11). MIT Sloan. Retrieved May 10, 2022, from https://mitsloan.mit.edu/faculty/directory/peter-m-senge

Peter Senge. (n.d.). New England Complex Systems Institute. Retrieved May 10, 2022, from https://necsi.edu/peter-senge

Ramage, M., & Shipp, K. (2020). *Systems Thinkers* (2nd ed. 2020 ed.). Springer.

Russ Ackoff. (n.d.). Global Association for Systems Thinking. Retrieved May 11, 2022, from https://545550393541748825.weebly.com/russ-ackoff.html

Schismogenesis | Psychology Wiki | Fandom. (n.d.). Psychology Wiki. Retrieved May 2, 2022, from https://psychology.fandom.com/wiki/Schismogenesis

Senge, P. M. (2006). *The Fifth Discipline: The Art & Practice of The Learning Organization* (Rev. ed.). Doubleday.

Servomechanisms | Encyclopedia.com. (2022, April 25). Encyclopedia. Retrieved May 9, 2022, from https://www.encyclopedia.com/science/encyclopedias-almanacs-transcripts-and-maps/servomechanisms-0

Shaping Forces - Margaret Mead: Human Nature and the Power of Culture | Exhibitions - Library of Congress. (n.d.). Library of Congress. Retrieved May 2, 2022, from https://www.loc.gov/exhibits/mead/mead-shaping.html

Systems and Systems Thinking | Encyclopedia.com. (n.d.). Encyclopedia of Science, Technology, and Ethics. Retrieved April 30, 2022, from https://www.encyclopedia.com/science/encyclopedias-almanacs-transcripts-and-maps/systems-and-systems-thinking

The Editors Of Encyclopaedia Britannica. (n.d.-a). *Margaret Mead | Biography, Contributions, Theory, Books, & Facts.* Encyclopedia Britannica. Retrieved May 2, 2022, from https://www.britannica.com/biography/Margaret-Mead

The Editors Of Encyclopaedia Britannica. (n.d.-b). *W. Edwards Deming | American statistician and educator.* Encyclopedia Britannica. Retrieved May 2, 2022, from https://www.britannica.com/biography/W-Edwards-Deming

The Editors Of Encyclopaedia Britannica. (2022, March 14). *Norbert Wiener | American mathematician.* Encyclopedia Britannica. Retrieved April 30, 2022, from https://www.britannica.com/biography/Norbert-Wiener

van Vliet, V. (2022, February 28). *Peter Senge.* Toolshero. Retrieved May 10, 2022, from https://www.toolshero.com/toolsheroes/peter-senge/

Veitch, J. S. (2007). *Peter Michael Senge (Printable)*. John Stephen Veitch. Retrieved May 10, 2022, from https://web.archive.org/web/20120426031344/http://www.openfuture.co.nz/petersenge.htm

von Bertalanffy, L. (1968). *General System Theory*. George Braziller.

Warren McCulloch. (n.d.). California State University Long Beach. Retrieved April 30, 2022, from https://home.csulb.edu/%7Ecwallis/artificialn/warren_mcculloch.html

Warren S. McCulloch – The Cybernetics Thought Collective: A History of Science and Technology Portal Project – U of I Library. (n.d.). University of Illinois. Retrieved April 30, 2022, from https://archives.library.illinois.edu/thought-collective/cyberneticians/warren-s-mcculloch/

Weckowicz, T. E. (1988). *Ludwig Von Bertalanffy's Contributions to Theoretical Psychology*. SpringerLink. Retrieved May 3, 2022, from https://link.springer.com/chapter/10.1007/978-1-4612-3902-4_25?error=cookies_not_supported&code=59e1859f-a11e-4a69-9a00-7ac775dc528d

Who was Norbert Wiener? Everything You Need to Know. (n.d.). The Famous People. Retrieved April 30, 2022, from https://www.thefamouspeople.com/profiles/norbert-wiener-8581.php

Wiener, N. (2021). *Cybernetics: Second Edition: Or the Control and Communication in the Animal and the Machine* (2nd ed.). Martino Fine Books.

Zeeman, A. (2021, November 2). *Jay Forrester*. Toolshero. Retrieved May 9, 2022, from https://www.toolshero.com/toolsheroes/jay-forrester/

Endnotes

[i] *Who was Norbert Wiener? Everything You Need to Know.* (n.d.). The Famous People. Retrieved April 30, 2022, from https://www.thefamouspeople.com/profiles/norbert-wiener-8581.php

[ii] The Editors Of Encyclopaedia Britannica. (2022, March 14). *Norbert Wiener | American mathematician.* Encyclopedia Britannica. Retrieved April 30, 2022, from https://www.britannica.com/biography/Norbert-Wiener

[iii] The Editors Of Encyclopaedia Britannica. (2022, March 14). *Norbert Wiener | American mathematician.* Encyclopedia Britannica. Retrieved April 30, 2022, from https://www.britannica.com/biography/Norbert-Wiener

[iv] The Editors Of Encyclopaedia Britannica. (2022, March 14). *Norbert Wiener | American mathematician.* Encyclopedia Britannica. Retrieved April 30, 2022, from

https://www.britannica.com/biography/Norbert-Wiener

[v] The Editors Of Encyclopaedia Britannica. (2022, March 14). *Norbert Wiener | American mathematician.* Encyclopedia Britannica. Retrieved April 30, 2022, from https://www.britannica.com/biography/Norbert-Wiener

[vi] The Editors Of Encyclopaedia Britannica. (2022, March 14). *Norbert Wiener | American mathematician.* Encyclopedia Britannica. Retrieved April 30, 2022, from https://www.britannica.com/biography/Norbert-Wiener

[vii] *Systems and Systems Thinking | Encyclopedia.com.* (n.d.). Encyclopedia of Science, Technology, and Ethics. Retrieved April 30, 2022, from https://www.encyclopedia.com/science/encyclopedias-almanacs-transcripts-and-maps/systems-and-systems-thinking

[viii] Wiener, N. (2021). *Cybernetics: Second Edition: Or the Control and Communication in the Animal and the Machine* (2nd ed.). Martino Fine Books.

[ix] Wiener, N. (2021). *Cybernetics: Second Edition: Or the Control and Communication in the Animal and the Machine* (2nd ed.). Martino Fine Books.

[x] Wiener, N. (2021). *Cybernetics: Second Edition: Or the Control and Communication in the Animal and the Machine* (2nd ed.). Martino Fine Books.

[xi] Abraham, T. (2016, October). *Rebel Genius*. The MIT Press. Retrieved April 30, 2022, from https://mitpress.mit.edu/books/rebel-genius

[xii] *Warren S. McCulloch – The Cybernetics Thought Collective: A History of Science and Technology Portal Project – U of I Library*. (n.d.). University of Illinois. Retrieved April 30, 2022, from https://archives.library.illinois.edu/thought-collective/cyberneticians/warren-s-mcculloch/

[xiii] *Warren McCulloch*. (n.d.). California State University Long Beach. Retrieved April 30, 2022, from https://home.csulb.edu/%7Ecwallis/artificialn/warren_mcculloch.html

[xiv] Abraham T. H. (2016). Modelling the mind: the case of Warren S. McCulloch. *CMAJ : Canadian Medical Association journal = journal de l'Association medicale canadienne, 188*(13), 974–975. Advance online publication. https://doi.org/10.1503/cmaj.160158

[xv] Abraham T. H. (2016). Modelling the mind: the case of Warren S. McCulloch. *CMAJ : Canadian Medical Association journal = journal de l'Association medicale canadienne, 188*(13), 974–975. Advance online publication. https://doi.org/10.1503/cmaj.160158

Abraham T. H. (2016). Modelling the mind: the case of Warren S. McCulloch. *CMAJ : Canadian Medical Association journal = journal de l'Association medicale canadienne, 188*(13), 974–975. Advance online publication. https://doi.org/10.1503/cmaj.160158

[xvi] Abraham T. H. (2016). Modelling the mind: the case of Warren S. McCulloch. *CMAJ : Canadian Medical Association journal = journal de l'Association medicale canadienne, 188*(13), 974–975. Advance online publication. https://doi.org/10.1503/cmaj.160158

[xvii] Marsalli, M. (2007). *McCulloch-Pitts Neurons*. The Mind Project. Retrieved April 30, 2022, from https://mind.ilstu.edu/curriculum/mcp_neurons/index.html

[xviii] Abraham T. H. (2016). Modelling the mind: the case of Warren S. McCulloch. *CMAJ : Canadian Medical Association journal = journal de l'Association medicale canadienne*, *188*(13), 974–975. Advance online publication. https://doi.org/10.1503/cmaj.160158

[xix] GoodTherapy Editor Team. (2018, May 3). *Gregory Bateson (1904–1980)*. The Good Therapy. Retrieved May 2, 2022, from https://www.goodtherapy.org/famous-psychologists/gregory-bateson.html

[xx] International Bateson Institute. (2018, November 14). *Gregory Bateson*. The International Bateson Institute. Retrieved May 2, 2022, from https://batesoninstitute.org/gregory-bateson/

[xxi] International Bateson Institute. (2018, November 14). *Gregory Bateson*. The International Bateson Institute. Retrieved May

2, 2022, from https://batesoninstitute.org/gregory-bateson/

[xxii] GoodTherapy Editor Team. (2018, May 3). *Gregory Bateson (1904–1980)*. The Good Therapy. Retrieved May 2, 2022, from https://www.goodtherapy.org/famous-psychologists/gregory-bateson.html

[xxiii] GoodTherapy Editor Team. (2018, May 3). *Gregory Bateson (1904–1980)*. The Good Therapy. Retrieved May 2, 2022, from https://www.goodtherapy.org/famous-psychologists/gregory-bateson.html

[xxiv] *Gregory Bateson - New World Encyclopedia.* (2008). New World Encyclopedia. Retrieved May 2, 2022, from https://www.newworldencyclopedia.org/entry/Gregory_Bateson

[xxv] Parks, T. (2018, December 15). *Gregory Bateson changed the way we think about changing ourselves | Aeon Essays*. Aeon. Retrieved May 2, 2022, from https://aeon.co/essays/gregory-bateson-changed-the-way-we-think-about-changing-ourselves

[xxvi] Ramage, M., & Shipp, K. (2020). *Systems Thinkers* (2nd ed. 2020 ed.). Springer.

[xxvii] *Schismogenesis | Psychology Wiki | Fandom.* (n.d.). Psychology Wiki. Retrieved May 2, 2022, from https://psychology.fandom.com/wiki/Schismogenesis

[xxviii] Ramage, M., & Shipp, K. (2020). *Systems Thinkers* (2nd ed. 2020 ed.). Springer.

[xxix] Ramage, M., & Shipp, K. (2020). *Systems Thinkers* (2nd ed. 2020 ed.). Springer.

[xxx] Kaiser, A. (2018, May 14). *Learning from the future meets Bateson's levels of learning | Emerald Insight.* Emerald. Retrieved May 2, 2022, from https://www.emerald.com/insight/content/doi/10.1108/TLO-06-2017-0065/full/html

[xxxi] Kaiser, A. (2018, May 14). *Learning from the future meets Bateson's levels of learning | Emerald Insight.* Emerald. Retrieved May 2, 2022, from https://www.emerald.com/insight/content/doi/10.1108/TLO-

06-2017-0065/full/html

[xxxii] Kaiser, A. (2018, May 14). *Learning from the future meets Bateson's levels of learning | Emerald Insight*. Emerald. Retrieved May 2, 2022, from https://www.emerald.com/insight/content/doi/10.1108/TLO-06-2017-0065/full/html

[xxxiii] Kaiser, A. (2018, May 14). *Learning from the future meets Bateson's levels of learning | Emerald Insight*. Emerald. Retrieved May 2, 2022, from https://www.emerald.com/insight/content/doi/10.1108/TLO-06-2017-0065/full/html

[xxxiv] Ramage, M., & Shipp, K. (2020). *Systems Thinkers* (2nd ed. 2020 ed.). Springer.

[xxxv] Ramage, M., & Shipp, K. (2020). *Systems Thinkers* (2nd ed. 2020 ed.). Springer.

[xxxvi] *British Library*. (n.d.). The British Library. Retrieved May 2, 2022, from https://www.bl.uk/people/w-edwards-deming

[xxxvii] The Editors Of Encyclopaedia Britannica. (n.d.). *W. Edwards*

Deming | American statistician and educator. Encyclopedia Britannica. Retrieved May 2, 2022, from https://www.britannica.com/biography/W-Edwards-Deming

[xxxviii] *Edwards Deming*. (n.d.). Global Association for Systems Thinking. Retrieved May 2, 2022, from https://545550393541748825.weebly.com/edwards-deming.html

[xxxix] *ASQ: About: W. Edwards Deming | ASQ*. (n.d.). ASQ. Retrieved May 2, 2022, from https://asq.org/about-asq/honorary-members/deming

[xl] *British Library*. (n.d.). The British Library. Retrieved May 2, 2022, from https://www.bl.uk/people/w-edwards-deming

[xli] Deming Collaboration. (2018, August 20). *Deming's 14 Points*. Retrieved May 2, 2022, from https://demingcollaboration.com/w-edwards-deming/demings-14-points/

[xlii] *British Library*. (n.d.). The British Library. Retrieved May 2, 2022, from https://www.bl.uk/people/w-edwards-deming

[xliii] *British Library*. (n.d.). The British Library. Retrieved May 2, 2022, from https://www.bl.uk/people/w-edwards-deming

[xliv] *British Library*. (n.d.). The British Library. Retrieved May 2, 2022, from https://www.bl.uk/people/w-edwards-deming

[xlv] *Shaping Forces - Margaret Mead: Human Nature and the Power of Culture | Exhibitions - Library of Congress*. (n.d.). Library of Congress. Retrieved May 2, 2022, from https://www.loc.gov/exhibits/mead/mead-shaping.html

[xlvi] *Shaping Forces - Margaret Mead: Human Nature and the Power of Culture | Exhibitions - Library of Congress*. (n.d.). Library of Congress. Retrieved May 2, 2022, from https://www.loc.gov/exhibits/mead/mead-shaping.html

[xlvii] *Shaping Forces - Margaret Mead: Human Nature and the Power of Culture | Exhibitions - Library of Congress*. (n.d.). Library of Congress. Retrieved May 2, 2022, from https://www.loc.gov/exhibits/mead/mead-shaping.html

[xlviii] *Shaping Forces - Margaret Mead: Human Nature and the Power of Culture | Exhibitions - Library of Congress*. (n.d.).

Library of Congress. Retrieved May 2, 2022, from https://www.loc.gov/exhibits/mead/mead-shaping.html

[xlix] History.com Editors. (2019, October 28). *Margaret Mead.* HISTORY. Retrieved May 2, 2022, from https://www.history.com/topics/womens-history/margaret-mead

[l] History.com Editors. (2019, October 28). *Margaret Mead.* HISTORY. Retrieved May 2, 2022, from https://www.history.com/topics/womens-history/margaret-mead

[li] History.com Editors. (2019, October 28). *Margaret Mead.* HISTORY. Retrieved May 2, 2022, from https://www.history.com/topics/womens-history/margaret-mead

[lii] B. (2021, March 29). *Margaret Mead.* Biography. Retrieved May 2, 2022, from https://www.biography.com/scholar/margaret-mead

[liii] History.com Editors. (2019, October 28). *Margaret Mead.*

HISTORY. Retrieved May 2, 2022, from https://www.history.com/topics/womens-history/margaret-mead

[liv] The Editors Of Encyclopaedia Britannica. (n.d.-a). *Margaret Mead | Biography, Contributions, Theory, Books, & Facts.* Encyclopedia Britannica. Retrieved May 2, 2022, from https://www.britannica.com/biography/Margaret-Mead

[lv] Boulding, R. J. (2018). *Elise Boulding: A Pioneer in Peace Research, Peacemaking, Feminism, Future Studies and the Family: From a Quaker Perspective (Pioneers in Arts, Humanities, Science, Engineering, Practice, 6)* (Softcover reprint of the original 1st ed. 2017 ed.). Springer.

[lvi] Ramage, M., & Shipp, K. (2020). *Systems Thinkers* (2nd ed. 2020 ed.). Springer.

[lvii] G. (n.d.). *Ludwig von Bertalanffy.* Environment and Ecology. Retrieved May 3, 2022, from http://environment-ecology.com/biographies/395-ludwig-von-bertalanffy.html

[lviii] G. (n.d.). *Ludwig von Bertalanffy.* Environment and Ecology.

Retrieved May 3, 2022, from http://environment-ecology.com/biographies/395-ludwig-von-bertalanffy.html

[lix] Ramage, M., & Shipp, K. (2020). *Systems Thinkers* (2nd ed. 2020 ed.). Springer.

[lx] Drack, Manfred; Apfalter, Wilfried; Pouvreau, David (11 March 2017). *"On the Making of a System Theory of Life: Paul A Weiss and Ludwig von Bertalanffy's Conceptual Connection"*. The Quarterly Review of Biology. *82 (4): 349–373. doi:10.1086/522810. PMC 2874664. PMID 18217527.*

[lxi] G. (n.d.). *Ludwig von Bertalanffy*. Environment and Ecology. Retrieved May 3, 2022, from http://environment-ecology.com/biographies/395-ludwig-von-bertalanffy.html

[lxii]

[lxiii] Drack, M., & Pouvreau, D. (2015). On the history of Ludwig von Bertalanffy's "General Systemology", and on its relationship

to cybernetics - part III: convergences and divergences. *International journal of general systems*, *44*(5), 523–571. https://doi.org/10.1080/03081079.2014.1000642

[lxiv] *Ludwig Von Bertalanffy - Work - Open Systems | Technology Trends*. (n.d.). Primidi. Retrieved May 3, 2022, from https://www.primidi.com/ludwig_von_bertalanffy/work/open_systems

[lxv] Weckowicz, T. E. (1988). *Ludwig Von Bertalanffy's Contributions to Theoretical Psychology*. SpringerLink. Retrieved May 3, 2022, from https://link.springer.com/chapter/10.1007/978-1-4612-3902-4_25?error=cookies_not_supported&code=59e1859f-a11e-4a69-9a00-7ac775dc528d

[lxvi] Drack, M., & Pouvreau, D. (2015). On the history of Ludwig von Bertalanffy's "General Systemology", and on its relationship to cybernetics - part III: convergences and divergences. *International journal of general systems*, *44*(5), 523–571. https://doi.org/10.1080/03081079.2014.1000642

[lxvii] Drack, M., & Pouvreau, D. (2015). On the history of Ludwig von Bertalanffy's "General Systemology", and on its relationship to cybernetics - part III: convergences and divergences. *International journal of general systems*, *44*(5), 523–571. https://doi.org/10.1080/03081079.2014.1000642

[lxviii] Drack, M., & Pouvreau, D. (2015). On the history of Ludwig von Bertalanffy's "General Systemology", and on its relationship to cybernetics - part III: convergences and divergences. *International journal of general systems*, *44*(5), 523–571. https://doi.org/10.1080/03081079.2014.1000642

[lxix] Ramage, M., & Shipp, K. (2020). *Systems Thinkers* (2nd ed. 2020 ed.). Springer.

[lxx] Drack, M., & Pouvreau, D. (2015). On the history of Ludwig von Bertalanffy's "General Systemology", and on its relationship to cybernetics - part III: convergences and divergences. *International journal of general systems*, *44*(5), 523–571. https://doi.org/10.1080/03081079.2014.1000642

[lxxi] Drack, M., & Pouvreau, D. (2015). On the history of Ludwig von Bertalanffy's "General Systemology", and on its relationship

to cybernetics - part III: convergences and divergences. *International journal of general systems*, *44*(5), 523–571. https://doi.org/10.1080/03081079.2014.1000642

lxxii Ramage, M., & Shipp, K. (2020). *Systems Thinkers* (2nd ed. 2020 ed.). Springer.

lxxiii Lee, J. A. N. (2016, November 16). *Computer Pioneers - Jay Wright Forrester*. IEEE Computer Society. Retrieved May 9, 2022, from https://history.computer.org/pioneers/forrester.html

lxxiv Zeeman, A. (2021, November 2). *Jay Forrester*. Toolshero. Retrieved May 9, 2022, from https://www.toolshero.com/toolsheroes/jay-forrester/

lxxv Zeeman, A. (2021, November 2). *Jay Forrester*. Toolshero. Retrieved May 9, 2022, from https://www.toolshero.com/toolsheroes/jay-forrester/

lxxvi Ramage, M., & Shipp, K. (2020). *Systems Thinkers* (2nd ed. 2020 ed.). Springer.

lxxvii Ramage, M., & Shipp, K. (2020). *Systems Thinkers* (2nd ed.

2020 ed.). Springer.

[lxxviii] I. (n.d.-b). *Forrester, Jay W.* INFORMS. Retrieved May 9, 2022, from https://www.informs.org/Explore/History-of-O.R.-Excellence/Biographical-Profiles/Forrester-Jay-W

[lxxix] I. (n.d.-b). *Forrester, Jay W.* INFORMS. Retrieved May 9, 2022, from https://www.informs.org/Explore/History-of-O.R.-Excellence/Biographical-Profiles/Forrester-Jay-W

[lxxx] Lee, J. A. N. (2016, November 16). *Computer Pioneers - Jay Wright Forrester.* IEEE Computer Society. Retrieved May 9, 2022, from https://history.computer.org/pioneers/forrester.html

[lxxxi] Zeeman, A. (2021, November 2). *Jay Forrester.* Toolshero. Retrieved May 9, 2022, from https://www.toolshero.com/toolsheroes/jay-forrester/

[lxxxii] I. (n.d.-b). *Forrester, Jay W.* INFORMS. Retrieved May 9, 2022, from https://www.informs.org/Explore/History-of-O.R.-Excellence/Biographical-Profiles/Forrester-Jay-W

[lxxxiii] Lee, H. L. (1997, April 15). *The Bullwhip Effect in Supply*

Chains. MIT Sloan Management Review. Retrieved May 9, 2022, from https://sloanreview.mit.edu/article/the-bullwhip-effect-in-supply-chains/

[lxxxiv] Ramage, M., & Shipp, K. (2020). *Systems Thinkers* (2nd ed. 2020 ed.). Springer.

[lxxxv] Ramage, M., & Shipp, K. (2020). *Systems Thinkers* (2nd ed. 2020 ed.). Springer.

[lxxxvi] I. (n.d.-b). *Forrester, Jay W*. INFORMS. Retrieved May 9, 2022, from https://www.informs.org/Explore/History-of-O.R.-Excellence/Biographical-Profiles/Forrester-Jay-W

[lxxxvii] Forrester, J. W. (2013). *Industrial Dynamics*. Martino Fine Books.

[lxxxviii] Lineausson, G. (2006). *School of Technology and Society DISSERTATION Literature Review on System Dynamics and Simulation*. University of Skovde. https://www.diva-portal.org/smash/get/diva2:1046998/FULLTEXT01.pdf

[lxxxix] Ramage, M., & Shipp, K. (2020). *Systems Thinkers* (2nd ed. 2020 ed.).

[xc] Springer. Forrester, J.W. Counterintuitive behavior of social systems. *Theor Decis* **2,** 109–140 (1971). https://doi.org/10.1007/BF00148991

[xci] Ramage, M., & Shipp, K. (2020). *Systems Thinkers* (2nd ed. 2020 ed.). Springer.

[xcii] Ramage, M., & Shipp, K. (2020). *Systems Thinkers* (2nd ed. 2020 ed.). Springer.

[xciii] *Peter M. Senge*. (2019, April 11). MIT Sloan. Retrieved May 10, 2022, from https://mitsloan.mit.edu/faculty/directory/peter-m-senge

[xciv] *Peter M. Senge*. (2019, April 11). MIT Sloan. Retrieved May 10, 2022, from https://mitsloan.mit.edu/faculty/directory/peter-m-senge

[xcv] van Vliet, V. (2022, February 28). *Peter Senge*. Toolshero. Retrieved May 10, 2022, from https://www.toolshero.com/toolsheroes/peter-senge/

[xcvi] van Vliet, V. (2022, February 28). *Peter Senge*. Toolshero. Retrieved May 10, 2022, from

https://www.toolshero.com/toolsheroes/peter-senge/

[xcvii] infed.org & infed.org. (2013, February 16). *Peter Senge and the learning organization – infed.org:* Infed. Retrieved May 10, 2022, from https://infed.org/mobi/peter-senge-and-the-learning-organization/

[xcviii] Clyburn, G. (2018, January 8). *The Fifth Discipline*. Carnegie Foundation for the Advancement of Teaching. Retrieved May 10, 2022, from https://www.carnegiefoundation.org/blog/the-fifth-discipline/

[xcix] infed.org & infed.org. (2013, February 16). *Peter Senge and the learning organization – infed.org:* Infed. Retrieved May 10, 2022, from https://infed.org/mobi/peter-senge-and-the-learning-organization/

[c] *Peter M. Senge*. (2019, April 11). MIT Sloan. Retrieved May 10, 2022, from https://mitsloan.mit.edu/faculty/directory/peter-m-senge

[ci] Veitch, J. S. (2007). *Peter Michael Senge (Printable)*. John Stephen

Veitch. Retrieved May 10, 2022, from https://web.archive.org/web/20120426031344/http://www.openfuture.co.nz/petersenge.htm

[cii] *Peter Senge.* (n.d.). New England Complex Systems Institute. Retrieved May 10, 2022, from https://necsi.edu/peter-senge

[ciii] *Peter Senge.* (n.d.). New England Complex Systems Institute. Retrieved May 10, 2022, from https://necsi.edu/peter-senge

[civ] Ramage, M., & Shipp, K. (2020). *Systems Thinkers* (2nd ed. 2020 ed.). Springer.

[cv] Grant, A. (2021). *Think Again: The Power of Knowing What You Don't Know.* Viking.

[cvi] Kang, P. (2021, January 24). *The Seven Learning Disabilities from The Fifth Discipline.* Peter Kang. Retrieved May 10, 2022, from https://www.peterkang.com/the-seven-learning-disabilities-from-the-fifth-discipline/

[cvii] Senge, P. M. (2006). *The Fifth Discipline: The Art & Practice of The Learning Organization* (Revised & Updated ed.). Doubleday.

[cviii] Ramage, M., & Shipp, K. (2020). *Systems Thinkers* (2nd ed. 2020 ed.). Springer.

[cix] Ramage, M., & Shipp, K. (2020). *Systems Thinkers* (2nd ed. 2020 ed.). Springer.

[cx] Senge, P. M. (2006). *The Fifth Discipline: The Art & Practice of The Learning Organization* (Revised & Updated ed.). Doubleday.

[cxi] Ramage, M., & Shipp, K. (2020). *Systems Thinkers* (2nd ed. 2020 ed.). Springer.

[cxii] I. (n.d.-b). *Ackoff, Russell L.* INFORMS. Retrieved May 11, 2022, from https://www.informs.org/Explore/History-of-O.R.-Excellence/Biographical-Profiles/Ackoff-Russell-L

[cxiii] *About Russell Ackoff.* (n.d.). Triarchy Press. Retrieved May 11, 2022, from https://www.triarchypress.net/ackoff.html

[cxiv] I. (n.d.-b). *Ackoff, Russell L.* INFORMS. Retrieved May 11, 2022, from https://www.informs.org/Explore/History-of-O.R.-Excellence/Biographical-Profiles/Ackoff-Russell-L

[cxv] *About Russell Ackoff.* (n.d.). Triarchy Press. Retrieved May 11, 2022, from https://www.triarchypress.net/ackoff.html

[cxvi] I. (n.d.-b). *Ackoff, Russell L.* INFORMS. Retrieved May 11, 2022, from https://www.informs.org/Explore/History-of-O.R.-Excellence/Biographical-Profiles/Ackoff-Russell-L

[cxvii] *Russ Ackoff.* (n.d.). Global Association for Systems Thinking. Retrieved May 11, 2022, from https://545550393541748825.weebly.com/russ-ackoff.html

[cxviii] Jaques, M. L. (1999). *Transformation and Redesign at the White House Communications Agency.* Quality Management Journal, Volume 6, Issue 3. Retrieved May 11, 2022, from https://curiouscat.net/pdfs/management/marchqmj99.pdf

[cxix] Ramage, M., & Shipp, K. (2020). *Systems Thinkers* (2nd ed. 2020 ed.). Springer.

[cxx] Kirby, M., & Rosenhead, J. (2005). IFORS Operational Research Hall of Fame: Russell L. Ackoff. *IFORS Operational Research Hall of Fame: Russell L. Ackoff, 12*(Intl. Trans. in Op. Res), 129–134.

[cxxi] *McGriff - Knowledge Base - Systems Theory*. (2001). Portland State University. Retrieved May 11, 2022, from https://web.archive.org/web/20070626112905/http://www.personal.psu.edu/sjm256/portfolio/kbase/Systems%26Change/systems.html

[cxxii] Kirby, M., & Rosenhead, J. (2005). IFORS Operational Research Hall of Fame: Russell L. Ackoff. *IFORS Operational Research Hall of Fame: Russell L. Ackoff, 12*(Intl. Trans. in Op. Res), 129–134.

[cxxiii] Ackoff, R. L. (1974). *Redesigning the future: a systems approach to societal problems* (Ex-library ed.). Wiley.

[cxxiv] Ramage, M., & Shipp, K. (2020). *Systems Thinkers* (2nd ed. 2020 ed.). Springer.

[cxxv] Ramage, M., & Shipp, K. (2020). *Systems Thinkers* (2nd ed. 2020 ed.). Springer.

[cxxvi] Ackoff, R. (2015, November 19). *A Lifetime of Systems Thinking*. The Systems Thinker. Retrieved May 11, 2022, from https://thesystemsthinker.com/a-lifetime-of-systems-

thinking/

[cxxvii] Checkland, P. (1999). *Systems Thinking, Systems Practice: Includes a 30-Year Retrospective* (1st ed.). Wiley.

[cxxviii] *Peter Checkland*. (n.d.). Lancaster University. Retrieved May 11, 2022, from https://www.lancaster.ac.uk/lums/people/peter-checkland

[cxxix] *Peter Checkland*. (2010). Academic Dictionaries and Encyclopedias. Retrieved May 11, 2022, from https://en-academic.com/dic.nsf/enwiki/251277

[cxxx] *Peter Checkland*. (n.d.). Lancaster University. Retrieved May 11, 2022, from https://www.lancaster.ac.uk/lums/people/peter-checkland

[cxxxi] *Peter Checkland*. (2010). Academic Dictionaries and Encyclopedias. Retrieved May 11, 2022, from https://en-academic.com/dic.nsf/enwiki/251277

[cxxxii] Ramage, M., & Shipp, K. (2020). *Systems Thinkers* (2nd ed. 2020 ed.). Springer.

[cxxxiii] *Peter Checkland*. (2015, June 18). Vanguard Consulting Ltd. Retrieved May 11, 2022, from https://beyondcommandandcontrol.com/library/whos-who-system-thinkers/peter-checkland/

[cxxxiv] Ramage, M., & Shipp, K. (2020). *Systems Thinkers* (2nd ed. 2020 ed.). Springer.

[cxxxv] Ramage, M., & Shipp, K. (2020). *Systems Thinkers* (2nd ed. 2020 ed.). Springer.

[cxxxvi] *Ervin Laszlo*. (2021, October 22). The Laszlo Institute. Retrieved May 11, 2022, from https://thelaszloinstitute.com/about/ervin-laszlo/

[cxxxvii] A. (2019, May 18). *What is the Akashic Field? by CJ Martes*. TOWARDS LIFE-KNOWLEDGE. Retrieved May 11, 2022, from https://bsahely.com/2016/11/13/what-is-the-akashic-field-by-cj-martes/

[cxxxviii] Laszlo, E. (2004). *Science and the Akashic Field: An Integral Theory of Everything* (Original ed.). Inner Traditions.

[cxxxix] *General Education Research Group*. (n.d.). The Darwin Project.

Retrieved May 11, 2022, from

https://www.thedarwinproject.com/gerg/gerg.html

[cxl] *Ervin Laszlo.* (n.d.-b). Ervin Laszlo. Retrieved May 11, 2022, from https://ervinlaszlobooks.com/author/ervin-laszlo

[cxli] *Ervin Laszlo.* (2021, October 22). The Laszlo Institute. Retrieved May 11, 2022, from https://thelaszloinstitute.com/about/ervin-laszlo/

[cxlii] Laszlo, E. (1972). *Introduction to Systems Philosophy: Toward a New Paradigm of Contemporary Thought* (1st ed.). Gordon and Breach.

[cxliii] von Bertalanffy, L. (1968). *General System Theory*. George Braziller.

[cxliv] Laszlo, E. (1972). *Introduction to Systems Philosophy: Toward a New Paradigm of Contemporary Thought* (1st ed.). Gordon and Breach.

[cxlv] von Bertalanffy, L. (1968). *General System Theory*. George Braziller.

Made in the USA
Columbia, SC
09 April 2024

F75ca213-208c-4257-9bd9-31cd29065b27R01

cxlvi Laszlo, E. (1972). *Introduction to Systems Philosophy: Toward a New Paradigm of Contemporary Thought* (1st ed.). Gordon and Breach.